T0295073

Energy IoT Architecture

From Theory to Practice

For a listing of recent titles in the
Artech House Power Engineering Library,
turn to the back of this book.

Energy IoT Architecture

From Theory to Practice

Stuart McCafferty

ARTECH
HOUSE

BOSTON | LONDON
artechhouse.com

Library of Congress Cataloging-in-Publication Data
A catalog record for this book is available from the U.S. Library of Congress.

British Library Cataloguing in Publication Data
A catalog record for this book is available from the British Library.

ISBN-13: 978-1-63081-969-9

Cover design by Andy Meaden Creative

Contents

3

Technical and Regulatory Barriers to the Energy Transformation   31

4

Energy IoT Reference Architecture Big Picture   43

5

Energy OT Domain: Evolving Towards a Neural Grid   55

6

Energy Business Systems (SaaS) Domain   81

7

Digital Energy Platform Services Domain: The Green Cloud   101

8

Mapping the IEEE 2030.5 Protocol to the Energy IoT Reference Architecture   127

9

Developing Energy IoT Rapid Solution Architectures   145

10

PNNL's Grid Architecture   167

11

The Path to Decarbonization Requires Integrated DER   173

Preface

"EnergyIoT Reference Architecture" was initially published in the first quarter of 2019 as a 7-part article series on Energy Central. It was one of Energy Central's most popular article series with over 25,000 views. The original series and subsequent publications won thought leadership honors for the Cleanie Awards for Community Contributor of the Year Gold Place Winner in 2021. These articles remain relevant and have generated conversations within utilities, digital cloud platform companies, regulators, and vendors.

Since publishing these articles, new distributed energy Internet of Things (IoT) solutions have begun to emerge, as well as new companies, and acceptance that the world of energy is experiencing unprecedented change. These new capabilities and business opportunities are being driven by climate change, viable integration of renewables and batteries into the electricity ecosystem, flexible loads, new market rules for more democratic participation and inclusion, transportation electrification, and financial support through new government grants, private equity, and venture capital investments.

Over the past several years, I have worked in the industry and have learned new tricks working with some extremely innovative companies. The energy transformation is happening for sure. Since the "EnergyIoT Reference Architecture" article series was published a few years ago, new technologies and approaches have emerged that were not included in the original article series.

This book takes the original concepts, structures, technologies, and practical application of IoT approaches to the next level. It addresses some of the lessons learned over the years since the articles' original publications, discusses minor yet nontrivial changes to the underlying architecture, and expands on the how of IoT approaches, modern digital cloud platforms, and communications abstraction that are driving innovation and a practical energy transformation that embraces distributed energy resources (DER) to create greater resilience, scalability, sustainability, and inclusion.

This is arguably the most exciting moment in the history of the electric power industry since the early days of Edison, Westinghouse, and Tesla. Change is happening, but it is challenging and sometimes even chaotic. A blueprint is needed to pragmatically implement a full energy transformation to easily support two-way power flow that directly addresses climate change issues to decarbonize the grid and build a much more inclusive, clean energy system. This system needs to be designed from the edge in, not from the utility enterprise out. This change is likely to come from new technology-savvy actors that implement smart, highly scalable IoT solutions that abstract the complexity of grid assets and energy systems to make integration simpler and more reliable. The change will almost certainly come from the edge.

Regardless of where the change comes from, change in the electric power industry is definitely in full bloom. However, the kind of change needed is more profound than the incremental and transitional solutions that continue to use the more than 100-year-old architecture and point-to-point technologies of the past. A new IoT reference architecture is the most practical way to achieve the transformational change that is needed.

If you are a student or a young professional looking for an exciting and profoundly impactful industry, the electric power industry is it. This book was written to help architects, students, utilities, policymakers, energy innovators, and people of all technical skill sets to recognize the various components of a successful energy IoT ecosystem as well as their dependencies and interactions with one another and how to apply them in their own designs.

Acknowledgments

When I was in the second grade, my teacher asked the class what we wanted to be when we grew up. The class was filled with future doctors, lawyers, astronauts, race car drivers, professional sports players, and millionaires. I very confidently answered that I would be a writer. Somehow, that early childhood dream evolved into a much different career path over the years. But that passion for writing has never faded and I have written hundreds of articles and documents throughout my career. When Artech House agreed to accept my proposal for this book, it fulfilled my lifelong ambition to be a published author. There are so many people that have helped me realize this dream. These acknowledgments can only begin to express my gratitude to these very special people.

I begin my acknowledgments with my family. My wife, Cindy, has been a rock and the primary breadwinner as I developed the book. She supported me while I dropped everything for 4 months to write it. Thank you, Cindy! Our kids, Kevin and Allie, have been cheerleaders throughout the entire process. My brother, Bob, also encouraged and supported me, reminding me that I talked about publishing a book when we were little.

My advisors at GridIntellect, Dave Kelly and Ken Fell, have provided daily advice and encouragement. My good friends and collaborators, Eamonn McCormick and David Forfia, spent hours on phone calls and web meetings discussing the energy IoT reference architecture to validate the concept and helped develop the original series published on Energy Central. David Forfia even coordinated my first presentation of the idea to the GridWise Architecture Council (GWAC) with the Council's feedback that I was onto something. Eamonn McCormick pushed me harder than anyone to write the book. My friends at Energy Central, especially Audra Drazga and Matt Chester, provided me with a publishing platform for numerous articles and support to get community feedback to help refine the concepts.

Several people reviewed my proposal and book draft to Artech House and provided constructive feedback. These industry experts included Dick Brooks, John Cooper, Paul Robinson, Ron Ambrosio, and James Mater. They truly shaped the book into its final format and encouraged Artech House to publish it. I cannot thank them enough, especially for their technical expertise and fact-checking efforts.

My publishing leads at Artech House, David Michelson and Natalie McGregor, stayed in regular contact and kept the ball moving. With this being my first book, they helped walk me through the process and found the right experts to help refine the book's technical storyline and editorial changes.

My editor, John Cooper, is an excellent writer himself and went above and beyond the call of duty. In fact, it seemed like John spent nearly as much time editing the book as I spent writing it! I couldn't have done it without him.

My friend, Jana Saddler, was nothing short of inspirational, providing constant encouragement when I was feeling burned out or had doubts about whether this was a worthy investment of time. Thank you, Jana.

Many people were involved in helping me get to a point where the book had the correct flow of information and encouraged me to stay the course. I am almost certainly missing some of the people involved, and I apologize to those I missed.

I hope that you, the reader, find the information presented in this book provocative and helpful in your work or classroom studies. I have spent over a decade studying, thinking, talking, and finally getting to this point. Hopefully this will translate into something you can apply in this disruptive energy transformation. Thank you for taking the time to read it. I sincerely appreciate you the most of all. Feel free to connect with me on LinkedIn or Energy Central.

1

Energy IoT: Get Your Head in the Cloud

Imagine a world where physical machines operating in the energy industry and the virtual software services of the Cloud were naturally linked through simple and reusable abstraction bridges. Imagine that these abstraction bridges could connect the virtual world with the physical world with the click of a mouse or, better yet, the two worlds could connect automatically with no human interaction. Imagine an electric power industry ecosystem that is easy to configure, leverages all utility and customer-owned energy resources to maximize energy efficiency, reduce greenhouse gas (GHG) emissions, and provide economic opportunities for utilities, vendors, and customers. We could make the leap from our siloed twentieth-century hardware-focused architectures to one that is silicon-based that not only transforms the way that our electric grid operates but also transforms the entire electric power industry and the way that customers interact with it. This positive change to the industry would ripple throughout world economies while maintaining safe, resilient, reliable, scalable, and affordable electricity delivery.

Apple changed the world with the introduction of the iPhone IoS at the Macworld Conference & Expo in 2007. This introduced a new "age of apps" that has catapulted yesterday's legacy phone to today's smartphone. IoS and Android operating systems created a mobile application bridge to allow innovators to create new value-added tools to help to simplify human lives. They created common development tools and application programming interfaces (API) to open the door to outside developers to enable common services supported by the IoS and Android operating systems. They simplified the way that users interface with smartphones in highly intuitive ways and exposed those capabilities to the development community. The explosive success of the smartphone has been revolutionary and created its own universe of opportunities, innovation, and growth that also supports new capabilities and efficiencies for the end user, making everyone winners.

The energy bridge will create similar success, providing digitalization, common APIs, and tools for utilities, vendors, customers, and innovators to connect energy assets to systems without the time-consuming, highly-skilled development techniques of the past.

It is time to think differently; it is time for the world of energy to change.

This chapter is focused on providing the reader with an understanding of three key factors that drive the need for change and the imminent disruption ahead for the electric power industry. These factors are happening today and are accelerating the need for immediate action. An energy transition is simply not enough. The industry needs to fundamentally transform, and the time to act is now.

A new era of smart energy optimization, innovation, resilience, clean green energy, reduced carbon emissions, and distributed control of physical grid assets is looming. The grid is becoming more distributed with dramatic growth in electric vehicles (EVs) and distributed energy resources (DERs), including renewables and batteries. Creating a bridge using energy-specific software services between the physical world of the grid and its connected assets and the world of apps will simplify integration and unlock the profound power of today's digital world. The digital energy services platform will propel the electric power industry from the electromechanical industrial revolution to a truly scalable digital world of the twenty-first century. The components needed to enable this cloud-based platform mostly already exist. We simply must assemble them and then recruit a user community to build and consume these services. Tethering the virtual world to the physical grid will be a dramatic step toward managing and controlling energy systems and devices for the better. It will also enable a new era of global innovation and opportunity globally that will have immense benefits for years to come. We call this innovation the digital energy services platform, or Green Cloud. This will digitize the world of energy and bring one of the most important industries in the world into the twenty-first century. Just like the iPhone propelled the human race towards a new mobile paradigm and the freedom to leave their desks without losing their ability to stay connected, the Green Cloud and the energy IoT reference architecture will propel us to a new world full of new opportunities and freedom.

As the driver of human civilization, energy has shaped human history and will have an even greater impact on the future. Over 100 years ago, there was an energy revolution driven by fossil fuels and centralized electricity generation. Another energy revolution is currently underway, propelled by growing societal mandates (both globally and locally) for clean energy, a steep decline in the costs of sustainable energy sources, an accelerating pace of technological innovation, and the continued expectation for delivery of energy to be safe, reliable, and affordable. The energy industry is in a period of impressive transformation, calling for a global change.

1.1   Driver #1: Societal Mandate for Clean Energy

The Intergovernmental Panel on Climate Change (IPCC) AR6 Climate Change 2021 report indicated that there are fewer than 10 years to "bend the curve" with atmospheric CO_2 and methane emissions. This means that the next 10 years will require a significant reduction in carbon and methane emissions on an annual basis. Policy changes at the global, national, and local levels are proceeding to address the observable impacts of a changing climate. The top two hottest years on record were 2016 and 2020, and 2018 was the fourth hottest year. Regular 100-year events are occurring every year now, with an increase in severe natural disasters as well.

Throughout the entire human existence, the Earth's average global temperature has been remarkably steady, fluctuating in a very narrow range of just 1°C (1.8°F). This is not to say that the

Earth has not experienced hotter and colder temperatures in its long history, but humanity has lived throughout an amazing climate stability. Yet, for the first time in human history, we are pushing beyond the limits of human experience and are undergoing unprecedented changes in temperature and climate, jumping a full degree in just 45 years and accelerating.

From Figures 1.1 and 1.2, atmospheric carbon and temperatures are at dangerous, never-seen-before levels in human history. Figure 1.3 shows the 1:1 correlation of atmospheric carbon or GHG levels to temperature from 1880 to 2020, the age of industrialization, where burning fossil fuels became the primary source of energizing cities, factories, cars, businesses, and homes.

Regardless of the cause being human or nature-related, we are in unprecedented territory with the amount of atmospheric CO_2 going back 400,000 years. Those periods of higher CO_2 concentrations correlate directly to increased global temperatures, jeopardizing our planet's ability to support life as we know it.

Nations, communities, and people have begun to take action. There are now widely accepted global agreements to reduce the impact of climate change by targeting the top sources of GHG emissions. The IPCC has escalated awareness of the Earth's disastrous predicament with climate change and the need to reduce our

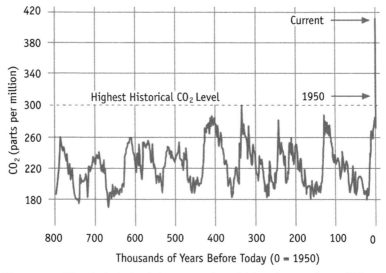

Figure 1.1 Historical CO_2 levels have never been higher. (Source: https://climate.nasa.gov/vital-signs/carbon-dioxide.)

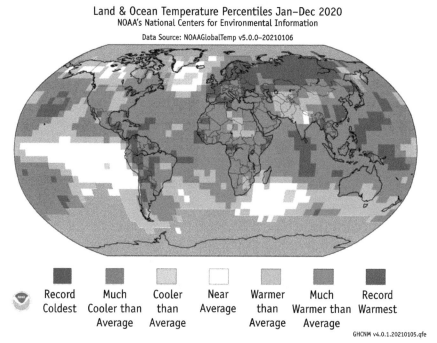

Figure 1.2   The Earth is heating up to dangerous levels never seen in human history.
(Source: National Oceanic and Atmospheric Administration (NOAA).)

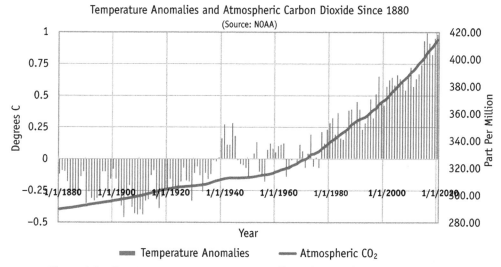

Figure 1.3   The 1880–2020 temperature anomalies and atmospheric CO_2 comparison.

collective carbon footprint to quickly move toward net-zero GHG emissions worldwide.

Transportation and electric power generation sectors are the leading sources of GHG emissions. In an effort to incentivize managing these industry GHG levels, at the national level, the United States created tax credits for renewable generation and electrified transportation options. In many global locations, cities are mandating 100% renewable requirements for electricity generation. More often, customers are willing to pay a premium for EVs and renewable energy electricity suppliers, rather than the less expensive traditional counterparts.

EPA data in Figures 1.4 and 1.5 show that the electric power industry is making a real impact in reducing its share of GHG emissions. According to the Sierra Club, 150 U.S. coal plants have been retired since 2010, and Figures 1.4 and 1.5 show a steep reduction in the electric power industry's overall contribution to GHG as a result. Since 2010, the electric power industry reduced its GHG contribution from 33% to 25%, while every other economic sector went up. Again, according to the EPA, since 2010, the electric power industry has actually reduced its CO_2 emissions from 2312 million metric tons (MMT) in 2010 to 1648 MMT in 2019, or nearly 29% reduction. All other economic sectors saw an overall MMT increase over the same time period.

So, you cannot make the claim that the electric power industry is not doing its part. It clearly illustrates that the industry can change and is clearly the most obvious and most impactful area to focus for GHG emission reduction and climate change mitigation.

However, the industry still needs to do more. This is especially true as the EV industry explodes onto the scene and threatens to completely disrupt the more than 100-year-old internal combustion engine (ICE) transportation industry. As many have said before, EVs fueled by dirty fossil fuel plants defeat the purpose. Transportation electrification is moving quickly, and the electric power industry must continue to embrace intermittent renewables, thoughtfully manage new intermittent loads, and provide nonintermittent, reliable electricity service to its customers.

In order to reach the Paris Agreement pledges and IPCC goals to keep global warming below 2°C (3.6°F), we need to decrease

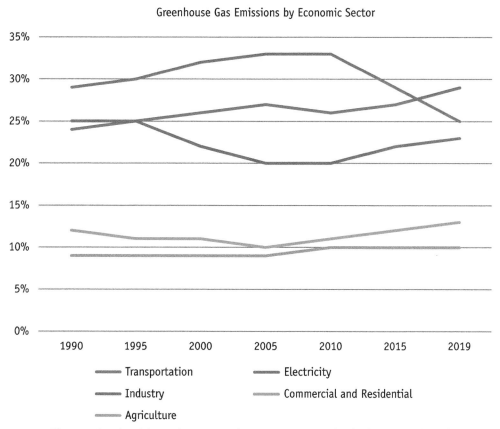

Figure 1.4 Electricity and transportation sectors account for the largest portion of GHG emissions. (Source: EPA.)

global GHG emissions by 3% per year. This is daunting, to say the least. However, that is exactly the rate at which the electric power industry has been on pace since 2010. This is primarily due to retiring coal plants and the increased use of solar and wind generation. The obvious high-impact emission reduction opportunities are quickly running out as the worst offenders, coal power plants, are rapidly retired and swapped for natural gas, wind, and solar. We will require new ways to create, manage, and reliably distribute clean electric power, which tends to be more distributed and less predictable than our legacy bulk power generation systems.

Clean energy is the practical path to mitigating climate change.

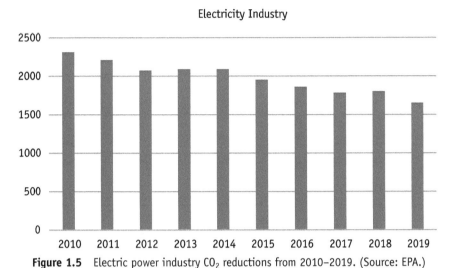

Figure 1.5 Electric power industry CO_2 reductions from 2010–2019. (Source: EPA.)

1.2 Driver #2: The Economic Advantage of Renewable Energy

Figure 1.6 was built from data from Lazard's Levelized Cost of Energy Analysis, Version 14.0 [1], from October 2020. The data are stunning and show a nearly tenfold reduction in the cost of solar photovoltaic (PV) crystalline from 2011 to 2020 to $37/MWh. Grid-

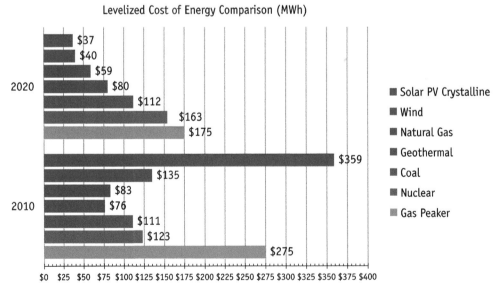

Figure 1.6 Lazard's Levelized Cost of Energy shows a dramatic reduction in renewable energy costs. (Source: Lazard's Levelized Cost of Energy Analysis Version 14.0.)

scale solar PV is now the cheapest form of electric power generation. Onshore wind is the second lowest cost at \$40/MWh and is down 70% since 2010.

Strictly from an economic point of view, the lower cost for renewable generation is driving investments in renewable energy deployments at the grid, distribution, and building owner levels as an overall cost savings mechanism for the utility and customer. Even if you remove the environmental benefits of renewables from consideration, the economic advantages of solar and wind are driving the replacement of traditional fossil generation with renewable sources across the globe. Also, wind and solar radiation are free; they are not commodities like coal and natural gas. Therefore, unlike fossil fuels whose prices fluctuate with market dynamics, the operational expenses (OPEX) of renewables are very predictable.

The National Renewable Energy Laboratory's (NREL) *Annual Technology Baseline* report [2] shows that this cost reduction trend is likely to continue. Strictly from an economic standpoint, the logic of investing in or operating traditional fossil fuel power plants is dubious, considering that utility-scale solar costs could decrease another 60% by 2050.

However, the Sun does not always shine, the wind does not always blow, and weather forecasts are not always accurate. The variability and predictable output decline at sunset of solar PV renewables will still require the mix of controllable loads or spinning mass generation to ensure balance across the grid and to manage rapid load ramp rates at sunset, famously described by the California Independent System Operator (CAISO) duck curve (see Figure 1.7).

Operational challenges resulting from higher penetration of renewable generation are a daily occurrence in Germany, Hawaii, California, Texas, and independent systems operators. Inevitably, the economics will continue to demonstrate a wider price difference. Concurrently, issues that are only being experienced by a few jurisdictions at present will rapidly expand to a broader base.

The economic advantage of renewable energy is currently confounded by utilities due to the operational challenges of controlling intermittent, more numerous, and smaller generation sources. These control issues are being solved in countries such as Australia by coordinating with utility and customer-owned assets

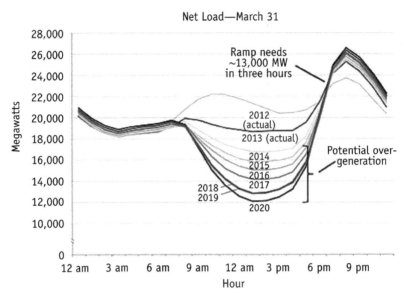

Figure 1.7   The CAISO duck curve shows the need for rapid power generation ramp-up after sundown due to renewable penetration.

using modern semantic model standards and message-based communications. They are providing blueprints for other utilities in all areas of the world. It will not be long before the economics of renewables are supported with other technologies such as energy storage that become standardized and commonplace. This will almost certainly reduce renewable generation costs even further and the dependency on bulk fossil generation plants will steadily decline.

Renewable energy is the cheapest form of electricity generation.

Because renewable energy is intermittent and not always predictable, energy storage is the "killer app" that makes renewable energy viable. The U.S. Energy Information Agency (EIA) and NREL use normalized cost values in their calculations, so the costs shown in Figures 1.8 and 1.9 do not necessarily line up with published cost projections. Regardless of which values are used, the cost of energy storage is dropping very rapidly, with the most rapid drop predicted from 2020 to 2025, with smaller cost reductions after that. This means that we are on the brink of an explosion of battery deployments as the costs quickly drive to a very affordable equilibrium.

Normalized Energy Capacity Costs (KWh)

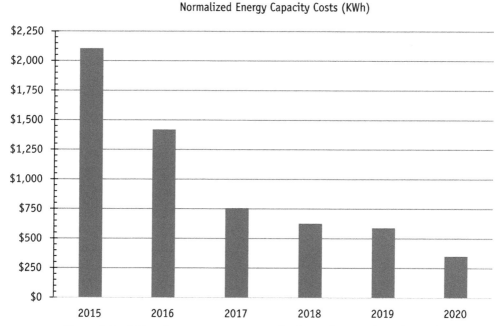

Figure 1.8   Grid-scale energy storage normalized capacity costs have dropped 72% since 2015. (Source: EIA.)

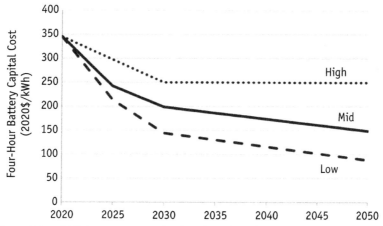

Figure 1.9   NREL battery cost projections for 4-hour lithium-ion systems. (Source: NREL.)

Renewable energy combined with energy storage delivers attractive economics and the necessary reliability to make it a viable alternative to bulk fossil fuel plant generation.

1.3   Driver #3: Technological Change Is Accelerating

Someone born in 1930 in the rural United States can probably still remember the day that an electric bulb replaced kerosene lamps. People born in the 1970s never knew a time when electricity was not widely available, but they can likely remember the times when storms or operational issues caused power losses for several hours or even days. Anyone born after 2000 cannot remember a time when light-emitting diode (LED) lightbulbs needed to be replaced, and they most likely will not remember a time before lights were verbally controlled through cell phones or other smart devices.

Technological change is occurring at a rapid rate, faster than previously seen throughout history. Ray Kurtzweil's concept of "Singularity" predicts that, by 2029, the artificial intelligence (AI) of machines will match that of humans [3]. Exponential advances and growth in technology have become the norm and humans quickly adapt and expect disruptive technologies to enhance their lives. The same will be true of the electric power industry, which is truly ripe for disruption.

The financial benefits of driving and meeting the increasing expectations of technology's ultimate consumers abound. Uber officially launched its mobile app in 2011. Although the Uber service was originally more expensive than cabs, it allowed riders to share costs in an easy, mobile transaction. By 2012, Uber created UberX, which was a cheaper ride-hailing service than taxis. At that time, Uber had a valuation of around $346 million [4]. Just 7 years later in April 2019, Uber reached a prepandemic market cap peak of $103 billion. Unfortunately, companies like Yellow Cab have essentially become obsolete due to antiquated thinking and a lack of technological foresight. Car rental companies and other ancillary agencies have experienced lost business based on users preferring the convenience of the Uber mobile application. Leveraging technology and establishing new customer expectations has not only simplified the user experience, but it has also helped to reduce costs, creating further differences between companies like Uber and car rental companies. The devastating effects to the incumbent industries are not lost on utilities and their traditional vendors. These companies must acknowledge that they cannot ignore technological change if they wish to remain relevant and profitable in this day and age.

Increased consumer expectations and technological growth are inevitable for the energy industry, on a scale and at a rate of change that is unlike anything ever experienced. As shown in Figure 1.10, technology companies such as Amazon have also seen enormous market cap growth as consumers rapidly adopt and recognize the convenience that disruptive technologies can provide. While disruption and change are somewhat simpler for companies like Uber, in the energy space, these changes must also support those "pesky laws of physics" [5] and policy rules set by local commissions.

Adapting to the disruptive forces ahead will require the same thought processes that industry-leading tech companies like Amazon, Google, Microsoft, Alibaba, and Uber have adopted to support dynamic, data-centric, elastic ecosystems within an adaptable regulatory structure that ensures that core principles are met. The opportunity for new players and new technologies in the energy space is enormous.

Technological change is accelerating. The change of consumer expectations is difficult to forecast not only because all energy decisions, like all politics, are truly local decisions, but also because

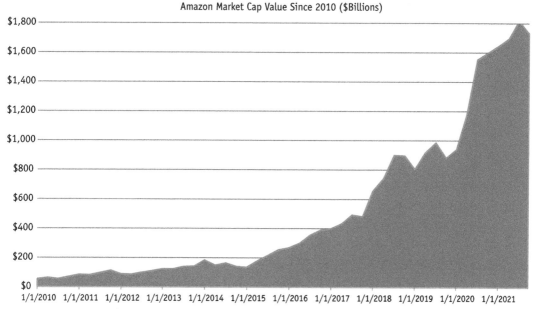

Figure 1.10 Amazon's market cap increased 32 times since 2010.

economic and demographic disruptions will occur suddenly and without sufficient lead time to respond.

Disruption is imminent for the electric power industry.

1.4   It Is Time to Act

Addressing any one of these drivers alone is challenging. Addressing all of them and those not yet contemplated requires the marriage of policy and technology. As the energy transformation accelerates, there will be new economic opportunities for utilities to provide new services and for innovators to create new capabilities. There will also be societal benefits for consumers to participate in markets, contribute to a greener climate, and utilize cost-effective renewable resources. Unlike anything experienced in the energy industry to date, the scope of the pending change will require not only different thinking from current central command and control system-centric architectures, but also policy changes needed to efficiently and economically make that transition.

Arguably, our greatest challenge is the short amount of time to be successful in this transformation to address these drivers.

The limits of what a top-down hierarchical grid (generation to transmission to distribution to load) and the hub-and-spoke, one-way electric power flow architectural paradigm can accommodate are being reached. Many utilities and other stakeholders are becoming increasingly more aware that an inherent limitation that we face is directly linked to how the current energy grid is constructed. The legacy power grid constrains our ability to transform the industry to meet the coming challenges, so the idea of an energy transition is truly more of an energy transformation. The electric power industry is struggling to meet today's challenges, especially in integrating and fully leveraging the bottom-up intermittent clean energy assets and technologies that are needed to scale to support millions of distributed assets. Yet no new fundamentally different and pragmatic reference architectures have been proposed that can support a bottom-up, scalable system while also accommodating existing legacy assets and systems.

Fundamental and practical architectural changes are required.

Going forward, we need to think differently to truly unlock the climate change, economic, and technology-driven solutions by:

- Enabling markets to spawn dramatically greater adoption of renewable energy, such as rooftop solar and democratized inclusion of anyone connected to the grid (see the GridWise Architecture Council's Transactive Energy Framework [6]).
- Creating scalable, extensible, flexible, data-centric architectures that support rapid change from primarily fossil-fueled energy generation to primarily clean energy generation.
- Developing policies and other incentives to drive the adoption of EVs, flexible loads, renewables, and energy storage solutions.

A transformed electric power industry includes these characteristics:

1. It is equitable and well-designed for small and large organizations; the industry must not only serve and preserve our investor-owned and public utilities, but also enable the creation of brand-new distribution services and energy-serving organizations. It must serve all stakeholders: municipalities, industrial customers, microgrids, and, most importantly, individual customers' businesses and households.

2. It meets requirements for societal and operational drivers, supporting local, regional, national, and global objectives, represented, with electric connectivity available to all, while also remaining safe, secure, reliable, and resilient.

3. It provides solutions for the critical deep electrification challenges facing the energy industry:
    (a) Decarbonized resource adequacy generation mix;
    (b) Solar and other renewable generation integration;
    (c) Energy storage integration;
    (d) Transportation electrification and integration with the grid;
    (e) Sustainable air conditioning and heating;
    (f) Load flexibility and controllability/coordination;
    (g) Resilience.

4. It is flexible. The global energy industry needs to enable multiple business models: integrated utility, distribution system operator (DSO) services, third-party virtual power plant (VPP) services, distributed energy resource (DER) aggregator market participation, cogeneration for large industrial, community microgrids, and other Energy (EaaS) as a Service utility and third-party models.

5. It is adaptable and extensible, supporting future-proof systems without requiring rearchitecting the system as new technologies and capabilities evolve over time:

   (a) Supports immediate and urgent needs of stakeholders such as enabling ubiquitous DER integration;

   (b) Provides pathways for existing and new innovative energy ecosystem service provider stakeholders such as Software as a Service (SaaS) providers, vendors, utilities, and DER aggregators;

   (c) Allows pathways towards the fantastical ideas not yet even imagined.

1.5 Disruption Ahead

Make no mistake; the electric industry is ripe for disruption. Without disruption and a complete rethinking of how electricity is generated and delivered, a crisis lies ahead. Considering the implications of where we are as an industry, the challenges that we face with scalability, siloed applications, stranded assets, DER integration, enabling edge intelligence, and simply retaining customers who seek to opt out and build their own microgrids, you recognize that change is imminent. As demonstrated by other industries, the Internet of Things (IoT) is a rapidly maturing and transformational technology that has very convincing advantages over today's top-down, system-centric, siloed, centralized architecture.

We need to address these forces head-on not only to keep electric power delivery reliable, safe, affordable, and adaptable to consumer needs, but also to enable policies to move to a greener industry and reduce the impacts of GHGs.

As you read this book, I hope that you feel the same sense of urgency that I do. I will make the case for why the electric power industry is the poster child for the IoT and why a three-layer IoT reference architecture is the practical and correct approach to solving some of our biggest industry challenges.

In the next chapter, I lay the groundwork for why the current centralized architecture cannot scale to support the rapidly evolving electric power industry ecosystem. In subsequent chapters, I dive into the energy IoT reference architecture and describe each component and its role in the new energy ecosystem.

References

[1]   lazard.com/media/451419/lazards-levelized-cost-of-energy-version-140. pdf, October 2020.

[2]   NREL Annual Technology Baseline, https://atb.nrel.gov/, 2020.

[3]   The Singularity is Near, https://en.wikipedia.org/wiki/The_Singularity_ Is_Near, Ray Kurtzweil, 2005.

[4]   Olsen, D., https://pitchbook.com/news/articles/uber-by-the-numbers-a-timeline-of-the-companys-funding-and-valuation-history, November 29, 2017.

[5]   Gunther, E., Smart Grid Subject Matter Expert and Evangelist, CTO EnerNex, 1959–2016.

[6]   GridWise Transactive Energy Framework Version 1.0, GridWise Architecture Council, 2015, https://www.gridwiseac.org/pdfs/te_framework_ report_pnnl-22946.pdf.

2

Architectural Challenges to the Energy Transformation

Change is difficult, especially for a regulated industry that over the past century has grown used to doing things essentially the same way they always have. Regulatory, architectural, and technology challenges must be overcome. Bidirectional electricity and distributed clean energy resources require fundamentally different approaches by the electric power industry to solve new challenges. The next two chapters address these challenges by describing the traditional architecture and technologies that the electric power industry has operated and why these new challenges can only be solved by radically different thinking.

In today's electric power ecosystem, adding new capabilities is tedious, time-consuming, and expensive. As an example, an advanced distribution management systems (ADMS) implementation can take years and tens of millions of dollars of investment. Even after integrating it with other utility systems and commissioning the ADMS solution, the amount of work and costs continue to be challenging. Maintaining even marginal situational awareness of the electrical networks in the utility's territory requires continuous, nontrivial vendor support and other OPEX as grid assets are retired or added and additional information technology (IT) systems are integrated.

Consider the contrast between a consumer solar power system and a new television set. When a homeowner purchases solar power, a PV system, even before installation, a system design from a certified electrician following the local building code process is required, then registration with the local power company, and waiting anywhere from several days to several weeks for an interconnect agreement.

However, the same homeowner purchasing a new television set simply needs to plug it in and connect it to the cable or satellite box. The cable/satellite company is indifferent to one more device added to its network. The system is up and running in minutes with no hassle or delay.

Why should the grid be so different from every other connected service? Is it that much more complex? Is it because the grid is a monopolistic ecosystem? Perhaps it is because grid architecture, policy, and processes have not evolved to enable a much more dynamic, democratic, and consumer choice-driven marketplace.

No one will argue that operating the grid is easy. Synchronization must be maintained to include correct voltage, amperage, and real and reactive power levels across the entire system to ensure power quality and reliability. However, we have also been doing this for over 100 years and have learned a lot about how to operate the grid. This learned operational knowledge and experience in moving electrons from generation to load can be leveraged today, but it is time to apply new IoT technologies to dramatically simplify adding and removing grid assets. We must envision and create a new ecosystem that enables the same type of plug-and-play concepts of the television set to how we add solar PV, flexible smart loads, battery, and EV assets. Let's take the solar PV system as an example in this new energy plug-and-play world. First, the consumer would purchase a solar PV system at his preferred home-improvement retailer; next he would either install it himself or pay someone to install it for him; and, when the consumer plugs in the newly connected system, it would:

1. Announce itself;
2. Describe itself;

3. Provision itself;

4. Commission itself.

Minutes after it was installed, the utility would know its identity, location, and capabilities. The power flow model would immediately be updated to include the new energy node, and the consumer could immediately leverage the ability to convert solar radiation into electrons and perhaps even bid that power into a local distribution market.

Theoretically, we could do this today; this is not science fiction. However, this is extremely unlikely given the existing architectural constructs and the legacy siloed utility systems. In Chapters 5 to 8, we propose a solution that leverages today's cloud and DevOps capabilities that companies such as Amazon, Google, Microsoft, and Red Hat have used for several years. This allows their systems and their customers' systems to elastically scale to massive dimensions in a simple, elegant, and practical way.

Our architectural challenges stem from how the grid has evolved over the past century. The grid represents multigenerational investments in a top-down architecture designed to deliver power from the central generation stations to the transmission grid to the distribution grid, all the way out to match energy consumer loads at the ends of the grid, the edge. Today's grid is an electrically synchronized complex machine with power flowing downhill from high voltage to low voltage across a vast network of wires, transformers, and switches, using a hub-and-spoke architecture.

Today's challenge is that this one-way load-following power flow model no longer aligns with how a transformed grid will work. Many would argue that it is just the opposite. The transformed grid will feature demand response-flexible smart loads that must follow vast numbers of distributed assets at the edge, including distributed generation, energy storage, electric vehicles, and as-yet-undeveloped devices. These new distributed assets are growing by leaps and bounds. They will continue to do so as DER prices drop, fossil fuel and electricity prices increase, policy changes turn adoption into law, and a societal call to action to address climate change becomes even more compelling. It is time to think differently.

2.1 Think of Everything as a Microgrid

If we can agree that the future grid transformation is a bottom-up model, then the next step is to consider each building block that makes up the grid as a microgrid: distribution networks, feeders, neighborhoods, buildings, and cars. These microgrid building blocks will make up the overall grid and manage themselves; they may have markets associated with them; and they will be able to run independently of a grid connection as an electrical island for periods of time. A single grid will become a nested microgrid system, a collective of microgrids that can operate both independently and in cooperation with other microgrids. This type of thinking is already occurring at progressive utilities such as San Diego Gas and Electric, where work is underway to pilot 10-MW batteries at some feeder locations to support microgrid capabilities, including islanding of the entire feeder. EVs are a great example of a mobile microgrid, an architectural component and actor in a larger microgrid building block when plugged in, and an islanded microgrid capable of operating independently for some period of time when unplugged. This distributed intelligence begins at the edge and propagates upwards to support larger and larger grid structures.

The Pacific Northwest National Laboratory (PNNL) developed their grid architecture concept with such a laminar layered approach. PNNL believes that each grid structure will have a coordination point to connect and interact in a friendly, orchestrated fashion with other grid structures. Obviously, this will not happen overnight, perhaps not in our lifetimes, but it is an interesting idea of where the electric power industry is likely heading over time.

2.1.1 What's Wrong with the Architecture We Have Now?

Even without taking the mental leap to "everything is a microgrid," today's grid is diverging from its original design. Changing from a one-way power flow model to a bidirectional flow is today's reality. Policy mandates accelerate such change, requiring rooftop solar on all new construction in California and renewable/clean energy targets in numerous states and even within the U.S. military. Yet most utilities still lack any situational awareness of assets located behind the meter (BTM). Two-way power caused by BTM assets that supply power back to the grid unchecked can result in over-generation and exceeding thermal limits of wires and transformers, causing damage or triggering protection equipment to trip offline. Other BTM assets that consume grid power unchecked, for instance, EVs, can cause similar consequences. Utilities are beginning to experience extremely challenging operational issues, such as the duck curve [1] (see Figure 1.7). As demand rapidly increases and solar PV assets stop producing electricity at sunset, unachievable power generation supply ramp rates may occur. Such looming safety and reliability threats can result that literally leave customers in the dark with the double impact of maintenance costs and lost revenue for the utility.

2.2   Challenges with Today's Electric Power Industry Architecture

The amount and density of renewables and energy storage assets necessary to reduce our reliance on fossil fuel generation is accelerating. Recognizing that such growth is here to stay, this section discusses some of the goals of a highly distributed clean energy ecosystem and the capacity of our current architecture to support those goals.

2.2.1   Reliable and Affordable Electricity

Historically, the top-down grid has been tasked to provide relatively affordable and reliable power. However, such stability may not remain with policies that assign a value or tax to atmospheric carbon contributors for their GHG impact. Electricity costs could rise dramatically if policymakers begin taxing carbon emissions. The Paris Agreement signed by 175 countries in 2016 has escalated awareness and agreement to address climate change head on through GHG reduction. In the bipartisan Climate Leadership Council's Four Pillars of a Carbon Dividends Plan, a modest neutral carbon tax as suggested by some Republicans led by ex-Secretary of State James Baker ("The Conservative Case for Carbon Dividends") of $40/ton of CO_2 could nearly double the cost of wholesale energy.

2.2.2   Fair and Equitable for Large and Small Alike

The current architecture is not working well for even the top utility companies, who struggle with high costs and have difficulty integrating distribution automation systems, DER, third-party aggregator VPPs, microgrids, EVs, and changes yet to come. Utilities in general struggle to integrate more than 20% renewables, meet policy objectives for clean power, and maintain grid reliability. Such challenges are compounded for smaller municipal public utilities and electric co-ops, which lack sufficiently large customer bases to fund the sophisticated distribution grid automation and situational awareness required for safe and reliable operations.

Policy, investment, and physical constraints limit choices for small producers and consumers. Limits on market participation, weather- and fire-related grid failures, and the risk of rising electricity costs actually incentivize some organizations and individuals to go off-grid by building their own microgrids.

2.2.3   Democratic, Secure, Trusted, Reliable, Resilient, and Safe

Following massive service disruptions in Texas, New Orleans, and Puerto Rico (to name a few), the grid's vulnerability to climate-related catastrophes has become ever more apparent. Further, the growing risk of cascading event cyberattacks from cybercriminals or foreign entities compounds such vulnerability, where sophisticated actors threaten both electric safety and reliability. The bottom-up hierarchy may make the grid prone to more attack surfaces and catastrophic events due to the larger number of assets, but these events can be isolated to keep damage as local as possible to reduce the opportunity for cascading consequences.

2.2.4   Decarbonization and Deep Electrification

The combination of digitalization, decarbonization, and deep electrification will have a wide impact on ending our reliance on GHG-emitting fossil fuels, expanding beyond the electric power industry to include all energy needs for transportation, agriculture, commercial and residential, and industry. Decarbonization and deep electrification will encompass lawn and garden tools, crop management, heating and cooling, trains, airplanes, and EVs, anything in life where traditional fossil fuels such as gasoline, diesel fuel, coal, or natural gas shift to clean electricity generation sources over time. Decarbonization and deep electrification are the two pillars in the race to zero carbon emissions to meet the Paris Agreement and IPCC goals and limit global warming to 2°C (3.6°F)

To achieve the panacea of a zero-carbon world, the thorny challenges associated with the electric power industry's current architecture must be overcome, as described below.

- *Transportation electrification:* Small numbers of EVs are currently manageable for grid operators, but higher numbers will require different strategies and technologies to manage unpredictable loads and protect the grid from overload. EVs represent something very new for grid operators: unpredictable mobile loads that can enter and exit various circuits and utility territories, complicating circuit-loading constraints as well as billing limitations. Cars, buses, trucks, trains, planes, farm equipment, ships and boats, and recreational vehicles are all forms of transportation that will be making the electrification transition, bringing along new challenges and opportunities. Even NASCAR has announced the introduction of an EV racecar series beginning in 2023.

- *Effective integration of renewable energy and energy storage:* Without energy storage, the intermittency of renewable energy will create grid reliability issues. Using today's top-down architecture, energy storage integration is costly, involving one-offs for every system. Utility command and control of DER assets uses archaic Supervisory Control and Data Acquisition (SCADA) systems that provide direct control by flipping registry bits, which takes time and effort to map, is prone to error, and makes troubleshooting complex.

- *Effective smart load management:* Smart, flexible loads are typically managed by aggregating assets of the same type or in the same general location. For instance, water heaters and thermostats are popular utility demand response (DR) solutions to provide load-side management, usually for peak load management. Traditional, dated utility DR programs generally use imprecise management techniques that send a control signal to numerous assets at a time. In most cases, these assets cannot be monitored (no situational awareness) and the grouping of assets cannot be changed easily. Further, thermostat programs can be gamed by the consumer by ramping the temperature setting up or down prior to an event. Also, thermostat programs can be misaligned with energy market needs. Reducing the Fahrenheit or Celsius temperature by a few degrees to bid into a market is misaligned with what utility customer programs really need: requests for kilowatt-hour (KWh) consumption changes rather than temperature value changes. Such imprecise legacy thermostat programs are likely to be replaced by more effective programs that coordinate with HVAC systems that actively manage load measured in KW and KWh in the years ahead.

2.2.5   Business Model Innovation

Business model innovation is one of the top challenges to utilities. The dependence on point-to-point, top-down operational systems and the regulated business model make it difficult to be innovative or think outside of the box to provide customers with new electric services. Policy and regulatory reform are required not only to safeguard the interests of stakeholders, but also to enable the ability of utilities to extend services beyond the meter. For instance, utilities may leverage DERs to provide higher reliability, guaranteed power quality, and better economy for its customers. The utility-owned microgrid business model is likely to gain traction, especially in the EV transportation sector where the utility can own the transportation microgrid, consume and store power when it is cheap or negatively priced, and provide grid services such as frequency and voltage regulation when needed to balance grid needs.

2.2.6   Sustainable Energy Future

The IPCC sets 2030 as the deadline to transform society to a sustainable net neutral posture and take us net-negative on emissions by 2050 at the latest. The grid is obviously a major player in reaching these goals. The current top-down reliance on bulk generation precludes a realistic pathway to a sustainable future, leaving a global shift to nuclear power as one of the most popular paths forward.

However, that path faces significant barriers: it would take years and billions of dollars to build, it does not ramp well with fluctuations in load and demand, and new plants may not be dispatched enough to recover their investment costs (i.e., become a "stranded investment"). The most feasible growth plan is to release exponential innovation via a coordinated, highly distributed, intelligent, decentralized, and flexible grid.

2.3   Utility's Siloed Systems

Dr. Jeff Taft, formerly the chief architect for Electric Grid Transformation at the Pacific Northwest National Laboratory, and Paul De Martini, the president at Newport Consulting, discussed the system-centric architecture issue head on in their groundbreaking "Sensing and Measurement Architecture for Grid Modernization" [2]. According to Taft and De Martini, the core problem is that grid systems, sensors, and data are hierarchical and rigidly bound to a specific utility system, forming disjoint data sets that cannot be leveraged by other applications. The hierarchical electric power architecture is system-centric rather than data-centric, which forces a siloed approach and leads to difficult and expensive systems integration challenges and orphaned data. The author used this example in Figure 2.1 to illustrate the challenge, showing both the application silo stacks as well as the complexity around data reuse and integration.

Taft and De Martini stressed that the vertical nature of essential system-centric structures leads naturally to the formation of silos. It is these silos that are the source of the fundamental limitations that drive the need for a new architecture. Applications such as Volt VAR Control, Circuit Fault Indicator, and Transformer Oil Analysis are configured as silos, with their own dedicated application system, communications systems, and data sets. The data silos that result lead to high integration and maintenance costs and little to no operational flexibility. It further limits the value of the data as analytics for optimization of grid services are not easily integrated with data from other external systems.

Taft and De Martini further identified that communications networks for grid sensors are generally hub-and-spoke, or, in the case of advanced metering infrastructure (AMI), local mesh to a

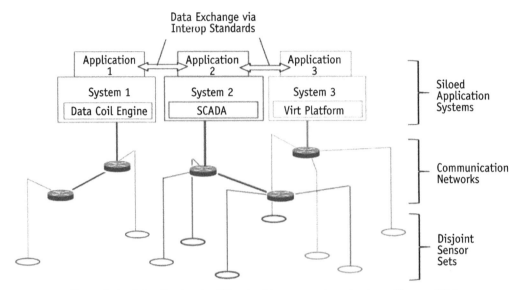

Figure 2.1　Data silos in a traditional grid sensor system structure. (Source: [2].)

hub-and-spoke backhaul (via cellular or substation: effectively, still hub-and-spoke). Such communication systems and data are siloed at the data collection head end and data storage systems. They have latency issues and are generally not scalable, and the utility grid centralized command and control systems (e.g., SCA-DA, AMI) are normally configured as point-to-point polling systems, requiring round-trip communications between grid assets and the centralized control system. Because of this polling type of communication, much of that traffic reports from grid assets that have no change in status, providing no additional value to the utility. This burdens utilities with high-bandwidth communication requirements that are expensive and must be extremely reliable to support the critical situational awareness of the utility and heavy communications traffic. Rather than using round-trip polling systems, a better solution would be to allow grid assets to automatically report status information only when changes occur and some threshold values are exceeded. This event-driven paradigm could dramatically reduce traffic, the amount of time series data that is stored, and the amount of computing power required to process and analyze real-time and stored data. Self-reporting grid assets could also improve interoperability by removing the siloed head-end systems by making data easily available in a common data lake repository for authorized systems.

The old paradigm of a top-down hierarchical, system-centric grid and vendor-driven solutions is simply incapable of moving the industry forward. Despite claims by some key industry vendors, most of these expensive solutions are not scalable and create high latency vertical silos, adding extremely high extensibility and integration costs. The resulting solutions generally underperform and are brittle and expensive to maintain, while delivering suboptimal business outcomes. The industry must find a new path forward to break out of this pattern. We must "un-silo" systems and data for better interoperability, move to event-driven communication paradigms, and operate from a data-centric perspective to allow authorized systems to leverage information for greater efficiencies and economics.

References

[1]   FlexibleResourcesHelpRenewables_FastFacts.pdf, California ISO, 2016.

[2]   Sensing and Measurement Architecture for Grid Modernization, https:// gridarchitecture.pnnl.gov/media/advanced/Sensor%20Networks%20for% 20Electric%20Power%20Systems.pdf, DOE, Taft and DeMartini.

3

Technical and Regulatory Barriers to the Energy Transformation

No doubt, the electric power industry is experiencing a period of unprecedented, exponential change and transformation. With numerous examples, Chapter 2 made it clear that the energy sector faces a series of systemic architectural challenges and opportunities that are undoubtedly already reshaping the industry in real time. Never before has the industry faced so much rapid change, and to make things even more challenging, that change is accelerating. This raises the question: Is the electric power industry on the edge of a transformational tipping point or not?

Despite such pressure and incentives for transformation, the electric power industry has been slow to change. Many in the industry do not yet see a point of no return and fall back to old habits and thought processes. What is even worse is that many still believe we can adapt to these changes by merely tweaking the existing system, leading one to ask: Why would people think this way?

3.1 Legacy Technology Mindset Is Based on Incrementalism

A key reason for this incremental mindset is the fact that, from a technology standpoint, the industry has survived, even thrived, based on incremental change for over 100 years. Despite world wars, industrialization, the rise of the internet, and even the introduction and growth of solar PV and wind power, the business model for regulated electric utilities has remained stable. As communications and power electronics have become faster and smarter over time, the electric grid has marginally improved in a stepwise fashion, and that has been accepted as enough, so far. The fundamental design of the modern power grid was laid down over 100 years ago by Edison, Tesla, Westinghouse, and Insull, and, beyond some deregulation and the introduction of regional bulk energy markets, not much has fundamentally changed since then. The majority of power is still generated centrally and delivered by a complex hub-and-spoke electricity transmission and distribution system to support customer loads.

However, more people now recognize the fundamental constraints and telltale signs, harbingers of an industry on the cusp of a historic transformation, one that is only set to accelerate. Our planet's very fate depends on an energy transformation that successfully and efficiently coordinates with distributed grid-connected assets, regardless of ownership, to reduce and possibly even eliminate our dependence on fossil fuels for electricity generation. This energy transformation will be instrumental in a global economic transition, ushering in a clean, efficient, resilient, customer-centric, and democratic future.

The centralized legacy system that served us well in the twentieth century is no longer sufficient to meet twenty-first century needs. We are moving to a distributed, large-scale, decarbonized, customer-centered, intelligent ecosystem. However, significant denial and skepticism challenging the necessity of a fundamental reshaping of the electric power industry persist. Too many in leadership positions remain blind to the large gap between our present condition and the urgent need for a future transformed industry and grid. The incrementalism culture is so strong that small, incremental changes are believed to be sufficient for the energy transformation, despite the profound changes and challenges now underway. A common industry mindset is that what worked yesterday will work just fine tomorrow.

There are many practical and good-intentioned reasons why the delivery of electricity has not changed for over a century.

Electricity has become a part of a modern Maslow's Hierarchy of Needs, joining air, food, water, and shelter as the most basic human needs. Electricity plays a central role in human progress and is subject to close government oversight. Over the years, a tremendous amount of capital has been invested incrementally in the current system, mostly in rate-based grid infrastructure assets established under regulatory depreciation rules, where such assets may be deemed operational for 20 years, but typically may remain in operation far longer, often until they fail. Considering the pace of technology change, it is incredible that industrial technology from 20, 40, or 60 years ago still remains in place. Throughout the grid, a "replace on failure" model still holds sway as a cost-effective strategy (i.e., "sweat the assets"). Incredibly, such "dumb" legacy assets remain in service performing their functions today just as they did before Neil Armstrong stepped out onto the moon. With so many assets lacking even basic digital intelligence, a centralized, direct control approach through a utility operations control room is still needed to coordinate legacy assets and manage the grid.

3.2   The Challenges of SCADA Systems

Utilities historically have managed the grid with 1940s-era telemetry systems that modulate and compress data into radio frequency (RF) signals, in order to squeeze as much information as possible into limited bandwidth. Legacy SCADA systems use telemetry technologies to send bits and bytes to assets to directly control the asset or read its registry values to obtain status information. As asset vendors employ different internal registries, SCADA systems must map to unique registry settings to ensure proper communication functionality. This time-consuming, cumbersome process requires significant maintenance and troubleshooting to resolve failures. Standards organizations have provided common registry interactions to provide some improvement, but such antiquated technology is simply a poor fit for today's intelligent assets. Further, direct asset control can actually cause damage or unexpected behavior in some circumstances. Inverters that support batteries, solar panel arrays, and EVs exemplify intelligent edge assets capable of "speaking" semantic languages and performing on request, no longer needing to operate as passive assets under the direction of a centralized SCADA system.

SCADA systems are also point-to-point, so data are not shared with other assets or edge devices. Modern IoT solutions have the capability to communicate locally with one another, which allows for distributed control among edge devices and gateways. Also, security challenges exist because most SCADA protocols communicate in clear text format and data can be intercepted or spoofed by bad actors.

SCADA systems continue to thrive in the electric power industry despite fundamental limitations. As mentioned previously, some of this has to do with the legacy infrastructure and legacy assets out in the field, waiting to fail. Another key contributor to perpetuating SCADA systems has to do with sunk costs, notably the enormous investments utilities have made in SCADA and supporting distribution and transmission coordination systems. Utilities have spent millions of dollars, in some cases hundreds of millions, on such centralized control solutions. While these systems worked well in the past, they are now reaching their capability limits as the number of assets grows and the top-down power flow model begins to transform into a bidirectional model.

3.3   The Challenges with Energy Management Systems (EMS), Distribution Management Systems (DMS), and Distributed Energy Resource Management Systems (DERMS)

Here we discuss the following systems:

- *EMS:* A utility enterprise SCADA-based centralized system for direct control and monitoring of assets on bulk power transmission networks. The term EMS is a bit overloaded because many smaller systems are named EMS, but for the purposes of this conversation, we are discussing the transmission operator's EMS.

- *DMS:* A utility enterprise SCADA-based centralized system for direct control and monitoring of assets on distribution networks.

- *DERMS:* A utility enterprise system owned and operated by the utility that enables the monitoring, management, coordination, and optimization of numerous DERs owned by the utility, its customers, or third-party aggregators to support grid operations and energy market participation.

- *VPP:* A virtualized grouping of DERs operated by a utility, a vendor, a customer, or a third-party aggregator (called DER aggregators) that enables the monitoring, management, coordination, and optimization of those grouped DERs for a variety of business purposes.

- *DER aggregator:* The actor providing VPP grid services.

Grid operators typically monitor the grid in real time with EMS and DMS and a centralized SCADA system to communicate with grid assets. These extremely sophisticated systems use visualization to provide operational grid status, including one-line diagrams and other graphic feedback with drill-down on the real-time status of individual assets or grid segments. Built-in logic may even provide operators suggestions and instructions to optimize grid behavior. EMS and DMS, often bundled together when purchased by the utility, can cost tens to hundreds of millions of dollars.

DERMS are relatively new utility enterprise systems developed to leverage DER assets to coordinate with traditional grid assets. Like other utility operations systems, DERMS are designed from the utility perspective from the utility enterprise systems to the edge, which is the opposite of how VPPs are designed. The basic idea with DERMS is to utilize DERs to support utility grid service needs when appropriate. DERMS "talks" to DER assets primarily through the SCADA system. However, most DERMS also have alternative communication pathways to the DER through the internet or cellular wireless telecom technologies, but this direct alternative communications pathway to individual DER is a rare mechanism utilized by DERMS. They also may communicate with third-party aggregators that can group several different DERs to support specific grid service needs. These third parties are often called DER aggregators or VPPs and provide connectivity through cloud-based services.

It is unfortunate that VPP, which implies a generation resource, is the name on which the electric power industry landed.

That certainly is one of the uses and a VPP can aggregate one or more small generation sources: renewables, gas and backup power gensets, and batteries. However, a VPP may also aggregate controllable loads to provide demand-side grid services for load flexibility or it may even aggregate EVs and EV-charging infrastructure. A more appropriate name, might be something like virtual grid asset aggregation, although that is not a particularly catchy name either.

The true value of EMS and DMS is their visualization and grid optimization capabilities, which will always be required. However, as the number of available grid assets grows, the communication pathways will need to expand beyond just SCADA. Currently, the DERMS provide some of that capability. DERMS, EMS, and DMS were all created basically as extensions of SCADA. Most people agree that DERMS will eventually evolve to be integrated with EMS and DMS, no longer a separate toolset. Current DERMS solutions are essentially patches until EMS and DMS can provide native support for DER coordination and management. However, it is likely that VPPs and third-party DER aggregators will continue to grow, in both number and capability, as DERs become dominant actors in providing grid services and even in participating in regional and local markets. Figure 3.1 provides a

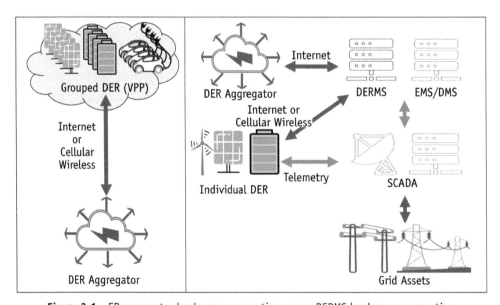

Figure 3.1   ER aggregator landscape perspective versus DERMS landscape perspective.

graphical representation and relationships of EMS, DMS, DERMS, SCADA, VPP, and DER aggregators.

3.4   The Challenges with Regulation

Given that these enormous system investments were not designed for the transformed energy ecosystem, it is easy to conclude that the regulatory model for a transformed electric power industry is fundamentally flawed. These systems, new 20 years ago, are no longer optimal. However, the sunk costs of existing infrastructure and supporting systems are recovered in rates paid by electricity consumers over time and only after local regulators representing those customers have approved the costs. Premature write-downs of these investments become stranded assets, which is bad business for both utilities and regulators.

Under the existing regulatory models in all states, utilities provide a rate case for grid upgrades to the local public utility commission (PUC). In essence, maintenance and upgrades are paid by rate payers upon PUC approval of construction projects needed to keep the grid running. PUC rate cases are designed to provide a guaranteed rate of return for the utility; these stable returns have made the monopoly utility industry an attractive investment on Wall Street. Rate cases have thus led to a utility preference for incremental large rate-based construction projects that can integrate with their existing legacy systems. Modern intelligent equipment such as grid-scale batteries and solar PV do not fit this incremental pattern, leaving utilities struggling to let go of their familiar SCADA mindset of direct control. As more of the grid-connected assets are owned not by the utility, but by customers and third parties, such attitudes are shifting, with progressive utilities starting to look beyond telemetry systems to semantic information models and IP-based communication technologies that communicate with distributed intelligent assets.

3.5   The Biggest Technology Challenge Is Scale

Centralized utility systems may work fine with small numbers of assets, but scaling of grid-connected assets to millions of devices will overwhelm communications channels. Legacy utility grid management systems will require enormous compute capabilities to even marginally operate the grid. However, there are solutions that can complement the utility's legacy systems and provide the scaling needs of the coming energy transformation. In the coming chapters, we will talk about an Industrial Internet of Things (IIoT) approach that leverages modern, elastic, highly scalable cloud technologies and reduces the dependence on centralized command and control systems and the amount of communications traffic needed to coordinate and manage large numbers of DER assets.

3.5.1 Digital Cloud Platforms Provide the Solution to Scaling Issues

An energy platform is a digital cloud-based architectural services layer that abstracts and autonomously scales to connect the real world of physical energy assets (operational technology (OT)) with the virtual energy business systems (energy enterprise), enabled by SaaS energy systems.

How will we meet the scaling needs of millions of grid-connected assets? The electric power industry will transform to be dominated by digital cloud platforms over the next decade. This transformation will primarily be driven by the IoT and the proliferation of large numbers of distributed renewables, energy storage, manageable loads, and EVs. As described by the U.S. Department of Energy (DoE), the electricity grid is an ultracomplex system of systems, so transformation will take time. However, we believe that accelerating drivers of change will realize a transformation to a high DER penetration digital platform-based industry within the next decade. This timeline is also in line with the need to achieve significant societal transformation goals for a customer-centric, decarbonized, distributed, clean digital energy, and electrified economy.

Before talking about the exciting entry of platforms to the pantheon of technology solutions, we should acknowledge that the term has often been overused. In the electric power industry and elsewhere, what some call platforms often fall short. For instance, SaaS vendors may refer to their solutions as platforms because APIs allow integration of their software to other systems, but that is insufficient. Rather the essential capability of platforms is the ability to perform dynamic scaling, what Amazon calls "elasticity," and that capability is only available in a handful of systems. Platforms distinguish themselves from SaaS solutions in many ways:

- Elasticity and scalability:
  - Virtual machines;
  - Container management and orchestration.
- Tools for identity management and security:
  - Identity management of individual and grouped system actors;
  - Role-based access;
  - Authentication and authorization;
  - Encryption;
  - Virus protection.
- Numerous data storage tools and microservices:
  - Structured data;
  - Unstructured data;
  - Digital ledger and smart contract technologies.
- Numerous workflow and data management solutions:
  - AI;
  - Workflow automation;
  - Low code development.
- Other DevOps, data, and workflow management microservices:
  - Adapters;
  - Digital twins;
  - IoT;
  - Messaging protocols and services;
  - Software development collaboration tools;
  - Version control;
  - Asset management;
  - Optimizers.

For more detail, Chapter 7 describes the energy business systems or energy enterprise, the conceptual architectural layer for energy SaaS solutions, and Chapter 8 focuses on the energy services domain, which provides energy services, the heart of the

Energy IoT reference architecture, and our description of a true energy platform.

3.6   Crossing the Technology Chasm

The true implications of the change to a digital platform industry have not yet fully sunk in. The 1990s bestseller *Crossing the Chasm, Marketing and Selling Disruptive Products to Mainstream Customers* by Geoffrey Moore described the vast chasm between early adopters and mainstream adopters. For the electric power industry, we need highly scalable, plug-and-play capabilities, a profound change from today's point-to-point energy solutions. Elastic digital platforms can scale dynamically to provide abstractions that enable simpler, more reliable ways to communicate and integrate smart edge devices. Platforms will provide the middleware architectural components that abstract the complex interactions between the OT hardware assets and the energy company's enterprise systems.

Security remains one of the primary concerns with transformation cited by utilities and regulators, who see a change from direct control of security from electric power industry professionals potentially introducing undue system risks and new vulnerabilities. Platform providers provide best-in-class security tools, processes, and resources as a core competency. A clear vision of the endgame must be clearly articulated if platforms are to be accepted by risk-averse electric power industry professionals.

The remaining chapters of this book will directly address such an endgame vision. Any structured transformation must begin with a clear vision, in contrast to the incremental patches so common among today's technology approaches. Applying incremental technology changes to existing electric power industry business models and architectures only delays progress, leading to increasing misdirection and failures. The profound changes now underway in the electric power industry mean that fundamental transformation has already begun. The visionary answer to the transformation challenge starts with new architecture, from which will flow changes (both incremental and transformative) to our use of technologies, to how markets operate, and to how policymakers interact with utilities, vendors, and customers, all to create fair and equitable policies to regulate the industry of the future.

3.7   Conclusion

The technologies and regulatory approaches of the past are insufficient to meet today's needs. Centralized command and control are outdated, given changes in asset ownership and technical capability that introduces self-management. Indeed, the challenges are numerous, complex, and vexing. For instance, the standard of 20-year amortization schedules can now outlive the useful life of most assets. The twin imperatives of energy transition and transformation put all status quo assumptions under the microscope.

The energy transition imperative to scale is without a doubt the biggest challenge, verging on a crisis. Consider that managing systems of systems that encompass diverse assets in the millions exclusively with a centralized SCADA providing exclusive direct control and communications is a nonstarter. A paradigm shift in how utilities and the rest of the electric power industry addresses the scaling crisis is imperative. The rational way to transform the industry and the climate change emergency must begin with an IIoT approach. The remaining chapters will discuss this new mindset and provide practical approaches to using an energy IoT reference architecture to point the electric power industry in the direction in which it needs to go.

4

Energy IoT Reference Architecture Big Picture

In previous chapters, we discussed why the energy industry is moving to a new paradigm. In the next four chapters, we will introduce the energy IoT reference architecture, which leverages modern architectural elements, such as virtualization, container-ization, orchestration, rich semantic information models, message buses, DevOps microservices, and the Cloud. This architecture:

- Is elastic;
- Is scalable;
- Is service-oriented and event-driven;
- Uses Internet Protocol (IP) addressing schemes;
- Will enable both edge computing and support legacy cen-tralized situational awareness computing needs;
- Is data-centric;
- Provides encrypted and authenticated communications, protects customer and data privacy, and provides role-based access to data;
- Uses rich semantics for communication (interoperability that is message-based instead of registry-based);
- Is extensible;

- Is designed to be future-proofed with the ability to adapt and extend API services;
- Provides fail-safe and redundant operations of edge assets;
- Can be located on site, in the Cloud, distributed on the grid edge, or in any hybrid combination of those options;
- Employs distributed hierarchical control and coordination;
- Is designed with expected communication losses.

The future grid must transform from a top-down to a bottom-up hierarchical paradigm and become much nimbler and more adaptable. The systems need to be able to scale from thousands to up to millions of assets. The technologies implemented need to become more aligned with IIoT methodologies, to include virtualization and containerization, standard information models and message buses, DevOps techniques, reusable microservices, and the flexibility and scalability provided by cloud-based environments. The systems should be designed to fully leverage modern mobile (e.g., 4G LTE and 5G) and wired communication technologies to deliver improved communication capabilities (but continue to autonomously run safely and efficiently when communication is interrupted).

New assets need to be discoverable and provision themselves to become active, dispatchable system elements quickly and simply. The system needs to be secure and resistant to local outages, regardless of the cause of the failure, be it the device, weather, or an intentional physical attack or cyberattack, and it must prevent cascading catastrophic events. Transmission and distribution systems must remain in balance by leveraging DER, including both distributed generation and flexible demand-side management (controllable loads). All customers and new businesses should be allowed to choose to participate in markets offering open access. It is important that this is done practically and with caution so that brown-field legacy assets and systems operate continuously during the transition. The energy IoT reference architecture that we describe in this book is designed with these fundamental guiding principles.

4.1   What Is Happening Here?

In the simplest terms, the energy industry is witnessing the integration of IT and OT. This transformation of the electric power industry is the epitome of an IIoT paradigm. Given the sheer scale, huge volume of data, opportunities for analytics and AI, and the amount and distance of distributed assets, the electric power industry is literally the poster child for IoT (as shown in Figure 4.1).

The energy IoT reference architecture in Figure 4.2 is a simple representation of some very complex system-of-system relationships. The figure easily represents the big picture, but it also allows each of the components to be considered individually, providing a clear comprehension of what role each component performs, the data generated, and how each element interacts with other ecosystem components. The architecture is event-driven and data-centric. The loosely coupled service-oriented architecture includes state-of-the-art security and access control techniques that reduce the likelihood of cascading events caused by natural disruptions, equipment failures, physical attacks to grid infrastructure, or cyberattacks.

Each grouped set of items within the two clouds and box are the domains of the energy IoT reference architecture. The two IT domains (energy business systems (SaaS) and digital energy services platform (Green Cloud)) include virtual components of the energy IoT ecosystem: software, middleware, core services, and data. The IT domains or portions thereof can exist in many places, including on site at utility data centers, hosted in the Cloud,

Electricity Industry—The "Poster Child" for the Internet of Things

- Electric Power Grid—"the biggest machine in the world!"
- Machine-to-machine communications and autonomous operations
- Big Data
- IP Addressable
- Virtualization
- Distributed
- Analytics and Artificial Intelligence

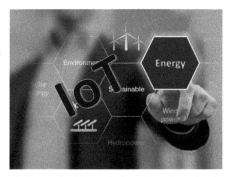

Figure 4.1   The electric power grid has all the IoT building blocks and scale to be the ultimate example of IoT.

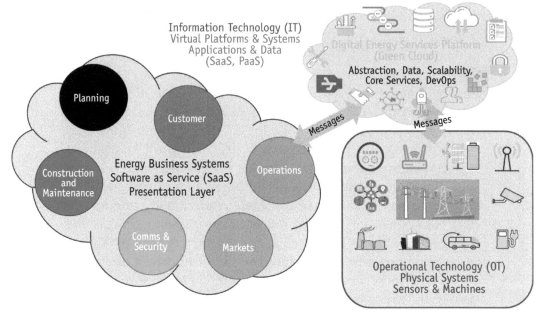

Figure 4.2 Energy IoT reference architecture.

or deployed on physical grid assets. The OT domain includes the grid, sensors, and machines. The physical OT and virtual IT domains need to work hand-in-hand.

4.2 The Energy IoT Stack View

Another way to consider the power of this architecture is through a technology stack architectural view.

Figure 4.3 shows the OT layer of physical grid assets at the bottom of the stack. The Platform as a Service (PaaS) or Green Cloud layer resides in the middle and abstracts the OT assets from the SaaS systems layer on top. The systems layer includes the businesses, consumers, and other stakeholders who ultimately use and operate the assets. The key point of this diagram is that most of what is represented already exists. The majority of core microservices in the PaaS middleware can be found today on any of the major Cloud vendors' sites. The energy-specific microservices are new services that must be developed and tuned specifically to support the operational and regionalized compliance requirements of

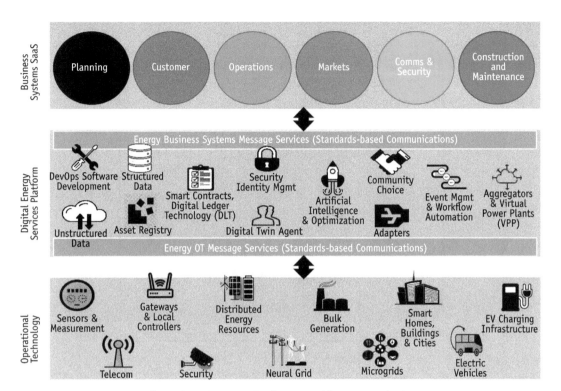

Figure 4.3 Energy IoT reference architecture stack view.

the electric power industry. Many building blocks are available today, and there are probably several new services still to be discovered. Transformation will not be simple or instantaneous, and inevitably there will be problems to be solved along the way. However, assembling an energy-specific Cloud platform services layer and developing the missing data-centric and associated communications abstraction services in Figure 4.3 can help to propel an industry-wide adoption of the energy IoT architecture ecosystem that we discuss throughout this book.

4.3 DER Device/OT Domain

The OT domain includes all the physical assets that make up the electric power grid including, but not limited to, wires, generators, substations, transformers, switches, remote terminal units, solar panels and inverters, energy storage, and any other potential

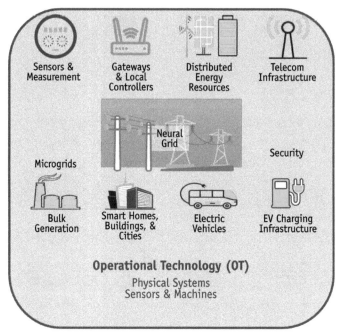

Figure 4.4 Energy IoT OT domain.

future unknown OT devices. As DER penetration grows, the grid will need to become nimbler and more adaptable (a neural grid) with assets that communicate and cooperate with one another to support stability and resilience.

The OT domain includes the following subdomains:

1. *Sensors and measurement:* A requirement for state information and situational awareness for devices, systems, and people.

2. *Bulk generation:* Large-scale generation assets needed to support baseline load capacity needs and ancillary services such as frequency and voltage regulation to meet reliability and power quality requirements.

3. *Telecom infrastructure:* Equipment and systems required to support communications with grid, utility, third parties, and customers.

4. *Aggregators and community choice:* System actors that provide grouped assets that can be used to support a wide variety of grid needs. Community choice aggregators are

a special type of aggregator that purchase wholesale and other local power to meet specific needs for their customers such as greener or less expensive electricity.

5. *Smart homes, buildings, and cities:* Necessary systems to coordinate generation, storage, EV, and flexible load assets to intelligently dispatch assets in homes, buildings, and cities.

6. *DERs:* Physical OT assets to include distributed generation, flexible/controllable loads, energy storage, EV infrastructure, and EVs.

7. *Security:* Physical security (e.g., cameras) and cybersecurity assets, systems, and processes used to protect the grid from bad actors.

8. *Electric transportation:* Systems and devices necessary to support electrified transportation.

9. *Neural grid:* Technologies that provide for safe, reliable, and nimble coordination with available grid assets and systems that use algorithms, machine learning, and other autonomous digital solutions to quickly reconfigure to adapt to planned or unplanned changes to the grid.

This domain will be discussed in more detail in Chapter 6.

4.4 Energy Business Systems (SaaS) Domain

The energy business systems SaaS domain shown in Figure 4.5 is a utility perspective of the software systems, devices, knowledge, and processes required to plan and operate the grid, broken out by subdomains:

1. *Planning:* Utility systems that forecast power requirements and physical grid infrastructure growth requirements for short-term and long-term planning.

2. *Construction and maintenance:* Utility systems that support physical construction projects, geolocation of grid devices, and the work orders to maintain and repair infrastructure.

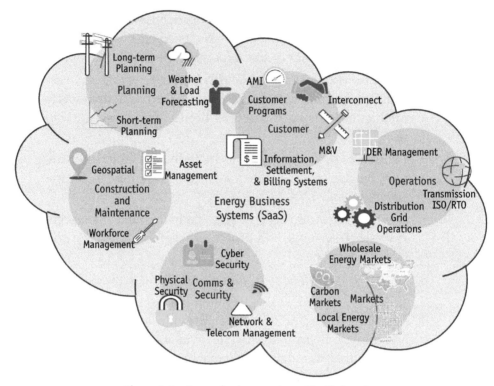

Figure 4.5   Energy business systems (SaaS) domain.

3. *Comms and security:* Utility software systems to support voice and data communications, along with physical security and cybersecurity services.

4. *Customer:* Utility systems designed to engage with customers, manage enrollment in different utility customer programs, provide interconnect services with customer assets, and determine settlement for services provided by the utility or the customer, and other customer information system (CIS) services that create customer bills and payments.

5. *Markets:* Utility systems that facilitate wholesale bulk transmission markets and future carbon and localized retail distribution markets.

6. *Operations:* Utility systems designed to provide command and control, situational awareness, and support safe and efficient operations of the transmission and distribution grid networks.

This domain will be discussed in more detail in Chapter 7.

4.5   Energy IoT Digital Energy Platform Services Domain (The Green Cloud)

The Energy IoT PaaS (also known as the Green Cloud) shown in
Figure 4.6 is the heart of this architecture. The Green Cloud services provide a common abstraction layer between the virtual and
physical domains, which dramatically simplifies interoperability
between physical devices and business software systems.

The Green Cloud enables a future plug-and-play electric
power industry ecosystem by facilitating:

- The customers' ability to connect DER to the electric distribution grid seamlessly and affordably;

- Enablement of customer choice, new energy offerings, and
  provisioning of grid services that meet the ever-evolving
  customer needs, wants, and preferences;

- Energy companies' ability to plan and build a network that
  can accommodate widespread interconnection of DER and
  use them to optimize the grid;

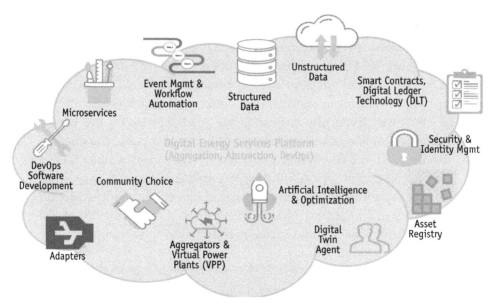

Figure 4.6   Digital energy service platform, the Green Cloud.

- Energy companies' ability to create and access grid services from DER by forecasting, monitoring, and controlling load and distributed generation;
- Integration of DER into the energy system, collecting and sharing data with authorized systems and people, and promoting innovation to decarbonize the electric distribution network;
- Enablement of the development of DER market opportunities for grid services that enable new interactions for customers and third parties;
- Enablement of third-party technology solutions safely, quickly, reliably, and affordably;
- Enablement of new third-party business models that integrate with the utility and other electricity service providers and consumers.

The energy IoT reference architecture was designed to support as-a-service business models that create interoperability and abstracted, streamlined communications between services, systems, and grid assets in an elegant, modern framework. The Green Cloud as-a-service APIs include energy-specific microservices, data repositories, and analytics to facilitate rapid prototyping and development for energy applications. Using the existing DevOps framework and microservices provided by the major Cloud vendors, these Cloud vendors already provide core techniques, scale, and virtualization necessary to support a DER-rich, bidirectional energy flow, dynamic, and adaptive grid. However, the Cloud vendors provide generic solutions and do not include the energy-specific additional services provided by the Green Cloud. These new Green Cloud services and framework leverage existing Cloud vendor capabilities and build energy data-centric models and repositories with supporting data services as the centerpiece of the architecture. This domain includes the following services:

1. *Smart contracts, digital ledger technology:* Decentralized technology that creates a common set of data that is managed by multiple participants and reduces the opportunity for hacking and changing data entries. The technology is trust-

less, so participants can trust the system but do not need to trust the other parties involved. It is immutable, meaning that the transaction service and its data cannot be changed once it has been set. All data are time-stamped and encrypted and the participants are anonymous. Because of the distributed nature of the data, all data sets must match, so it requires unanimous agreement for a transaction to be valid. This type of technology could be used in the electric power industry for a variety of applications, including transactive energy, settlement services, and perhaps even tracking "green" electrons.

2. *Structured data:* Relational information, such as time-series or grid configuration data, stored in relational databases and data lakes.

3. *Unstructured data:* Provide storage and retrieval of documents, XML, pictures, 3-D renderings, and other raw data files.

4. *Digital twin agent:* A virtual representation of an asset or group of assets that abstract and simplify communications, provide data persistence, manage events, communicate with other parts of the ecosystem, and optionally provide emulation capabilities or 3-D models for those assets.

5. *Security and identity management:* Tools that support management of actors within the system for authentication and authorization, provide encryption for communications and data, virus protection, and role-based privileges to elements within the ecosystem.

6. *Source code management:* Collaborative tools for software developers to manage, store, modify, and extend source code as well as providing configuration management and version control. Tools like GitHub are popular source code management systems.

7. *Orchestration:* Provide, execute, and choreograph workflows and processes.

8. *Microservices:* Reusable Cloud API services that create "black boxes" for developers to abstract complexity through simpler interfaces to create applications

faster without requiring detailed knowledge of how the microservice works.

9. *AI and analytics:* Computer-based digital tools used to locate anomalies, optimize processes and outcomes, and may even provide decision-making guidance on how to address issues that that have been uncovered.

This domain will be discussed in more detail in Chapter 8.

4.6   Conclusion

The energy IoT reference architecture is an entirely unique architectural construct from the hub-and-spoke architecture of today. The grid of the future is coming more quickly than one might think. New methods and technologies to scale and support rapid grid configuration changes are needed. The energy IoT reference architecture, designed to utilize today's most advanced technologies, uses and extends existing Cloud services with energy-specific microservice APIs, is standards-driven, and abstracts the complexity of communicating and managing grid assets, both in front and behind the meter. The architecture was designed with future-proofing in mind and the ability to evolve as technology changes occur, new asset types and API services emerge, and the growth of DERs overwhelms our legacy systems.

The key takeaway is that the energy IoT reference architecture was designed to become a plug-and-play digital energy services platform or Green Cloud that allows rapid integration of DER devices, then bridging those devices with a new world of SaaS software applications and innovative new business models. As the effects of climate change become more severe, the energy IoT reference architecture provides a practical and very deliberate design to achieve a sustainable, rapid, clean, and effective energy transformation.

This cannot be done by one company alone. It will take a multitude of people with similar passion and focus to solve the electric grid challenges we are currently facing and create the inclusive, decarbonized, plug-and-play grid of tomorrow.

5

Energy OT Domain: Evolving Towards a Neural Grid

This chapter focuses on the physical OT domain within the model, which is highlighted in the red box on the bottom right corner of Figure 5.1. The OT domain envisioned:

- Has intelligence, a neural grid.
- Is nimble and adaptive.
- Facilitates rapid interconnect and provisioning of conventional grid assets and DER assets.
- Accelerates electrification of the transportation industry.
- Provides reliable, resilient, and safe electricity delivery.
- Is completely interoperable.
- Grid needs can be met using the appropriate available grid assets regardless of ownership.
- Is a bottom-up hierarchical ecosystem built from local distribution network edge atoms to transmission level systems.
- Is secure in both the physical domains and the cyberdomains: authentication, authorization, encryption.
- Is economical.
- Is inclusive.
- Accommodates opportunities for innovation and new services.

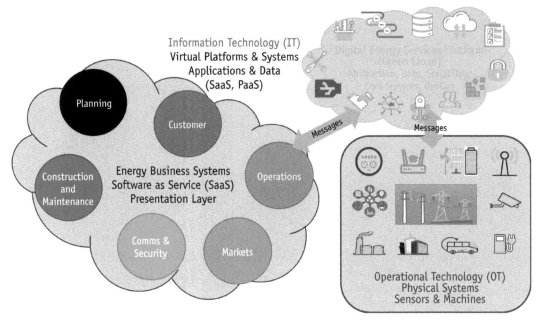

Figure 5.1   Energy IoT reference architecture with the OT layer at bottom right.

5.1   The Neural Grid

At the center of the OT domain is a self-aware, intelligent neural grid. This concept of a neural grid aligns with DoE PNNL's grid architecture model (see Chapter 10) where smaller grid structures (as an example, a distribution feeder or a circuit within a feeder) can be organized into larger layered or laminar grid structure hierarchies. Each of the smaller grid structures can manage themselves and support larger hierarchies through coordination points that allow them to interoperate with a parent and other child structures within the hierarchy. This distributed intelligence concept was also discussed in Section 2.1 and is the idea of bottom-up layered intelligence.

In 2007, the National Energy Technology Laboratory (NETL) published *Modern Grid Benefits*, which included 7 principal characteristics of what we now call the smart grid. One of those characteristics was the concept of self-healing. In reality, today's smart grid is already evolving towards a neural grid. The neural grid evolution will not only support the concept of self-healing, but over time could also include self-organizing logic with the additional capability of optimizing grid operations using the structures and assets available. This will require new power electronics technologies, control/coordination standards and software services, and intelligent coordination points across the physical grid landscape. These technological advances will allow individual grid structures to island themselves and manage the substructures and assets within its hierarchy, operating independently for long periods of time.

This kind of advanced adaptability is likely many years away from realization, but conceptually, this is the type of grid intelligence envisioned within the energy IoT reference architecture's neural grid and that the architecture is designed to support. Further, the architecture was designed to allow participation by a variety of different actors when and however they can. The energy services platform (Green Cloud) described in Chapter 7 will provide the abstraction necessary to support these actors to dramatically simplify and accelerate integrating IT systems with OT grid assets. The concept of a digital twin agent, the mechanism that helps enable this hierarchical functionality, will be discussed in Chapter 7.

5.2   Sensors and Measurement

A key aspect of the neural grid is self-awareness, and additional sensors with low-latency measurement feedback will be required for the neural grid to understand its current state information. Granular grid measurements will be necessary to support each grid structure to allow the structure to react to current conditions, get help from other parts of the grid, and self-heal or isolate when problems occur. Sensor measurements such as phase angles and real and reactive power will provide grid structure coordination points with information to allow them to maintain acceptable operating ranges for balancing and power quality. Digital phasor measurement units (PMUs) are rapidly growing beyond just transmission systems and are being deployed at congestion points on distribution networks; PMU usage will continue to accelerate as prices drop, DER penetration increases, and the ability to manage the grid at more granular levels increases.

Current and forecasted weather conditions are also critical measurements to grid operations when renewable energy assets

such as PV solar and wind are involved. Solar radiation, temperature, precipitation, and wind speed and direction are required to
provide much more granular hour-ahead, day-ahead, and week-
ahead planning and forecasting input. These measurements provide inputs that factor into load and generation forecasts that also
ultimately determine electricity prices in capacity and ancillary
electricity markets.

5.3 Gateways and Local Controllers

Gateways and controllers are often merged into a single piece of
industrial computing equipment. A local controller provides localized command and control of the assets within its purview. The
local controller may operate based on commands or requests from
a centralized system (most likely) or it may work independently
in nongrid-connected or islanded modes of operation. Microgrid
controllers are good examples of local controllers, but microgrid
applications are not the only application of local controllers.

Gateways provide abstraction of local energy assets to centralized systems. They perform protocol translations to allow different communications languages to work with one another. For
instance, a utility system may talk to a gateway using the distributed network protocol 3 (DNP3) via the SCADA telemetry system.
The gateway can then translate those DNP3 messages into IEEE
2030.5, Modbus, OPC-UA, or whatever protocol an individual local asset uses for communication. Many gateways are emerging
that also include aggregation and control capabilities to allow
centralized systems to communicate with the gateway as a single
point of aggregated capabilities. The gateway receives the central
system messages of what the central system wants from the aggregated assets and then makes decisions on how to manage those
local assets to provide the needed services.

5.4 DERs

DERs are arguably the biggest technology catalyst to require different architectural approaches than those used for the past century. For the purposes of this book, the author considers the Na-

tional Association of Regulatory Utility Commissioners' (NARUC) definition of DER as most applicable [1]:

> A resource sited close to customers that can provide all or some of their immediate electric and power needs and can also be used by the system to either reduce demand (such as energy efficiency) or provide supply to satisfy the energy, capacity, or ancillary service needs of the distribution grid. The resources, if providing electricity or thermal energy, are small in scale, connected to the distribution system, and close to load. Examples of different types of DER include solar PV, wind, Combined Heat and Power (CHP), energy storage, DR, EVs, microgrids, and energy efficiency (EE).

A DER asset is typically described as a small grid-connected device that provides distributed generation, a controllable load, or energy storage, normally located on the distribution network.

The sheer number and rapid adoption of DERs are creating highly complex challenges for the top-down grid architecture including:

- Daylight hours overgeneration from solar PV devices with extremely high ramp rates in the afternoon when workers return to their homes as the Sun goes down (i.e., the duck curve).
- Limited or completely missing situational awareness of nonutility assets located behind the meter.
- Interconnect agreements and utility processes that require in-person inspections and approvals with long interconnection queues.
- Heightened demand for further distributed and more granular grid and sensor measurements (e.g., PMUs, weather).
- Time-consuming and high costs to install and integrate DERMS with other utility operational systems to coordinate and manage DER. There are no enterprise DERMS that can coordinate and manage all DER assets that include those that are nonutility-owned and behind-the-meter.

- Communication and supporting systems to integrate DERs are costly.
- Maintaining DERMS and supporting systems are time-consuming and expensive.
- Closed or proprietary vendor semantics or protocols, the sheer number of existing equipment standards and variations, and the large number and variety of DER assets that need to be coordinated with can create brittle integrations that are easily broken when software changes or upgrades are needed.
- Inadequate network and DER asset information that result in short-term and long-term grid needs' forecasting and planning errors.
- Electrical network restoration after storms or emergency load shed events that lack adequate situational awareness or coordination with local DER assets may result in large voltage or frequency swings that can damage equipment or create additional problems for customers and utilities.
- Difficulty sustaining and operating power flow models due to erroneous inputs to the state estimator and load adaptation.
- Ability to organize and manage DERs to support dynamic response to faults, reactive power requirements, voltage or frequency regulation needs, and other grid events.
- Distribution-level congestion management.

Even though DERs have been challenging for utilities to date, they also introduce exciting opportunities that include:

- The ability to rapidly decarbonize the grid and reduce our fossil fuel power generation dependence;
- Decreasing customer electricity costs;
- Creating local inclusive distribution markets and other DSO grid management capabilities;
- Providing a variety of new local customer grid services that provide better power quality, greener energy, resilience, and lower costs;

• Increasing reliability and resilience that also minimize restoration times when failures occur.

The current top-down hub-and-spoke architecture fails to address these challenges or take advantage of the numerous opportunities cited above. The traditional top-down architecture currently in use cannot scale to support the local, national, and international GHG reduction policies that incentivize or require much larger utility and customer adoption of renewables, energy storage, and EVs. The Green Cloud digital energy services platform domain in the energy IoT reference architecture presents an abstraction layer intended to simplify and speed up new DER interconnection and standardize integration with other grid assets and systems (including those in central operations centers). This will be discussed in more detail in Chapter 7.

5.4.1   Energy Storage DER

Energy storage is a special type of DER that merits further discussion. Because batteries have the ability to consume, supply, and store power, they have emerged as the "killer app" for energy systems. Perhaps the biggest argument that utilities and policy makers had in the recent past with renewables is the intermittent nature of generation. The need to store energy in a high-penetration DER environment cannot be overstated. Overgeneration from PV solar and wind generation plants is wasted power if it cannot be used to support other system loads or stored. Challenges with CAISO's famous duck curve where PV overgeneration can result in near-zero or negative electricity prices and steep ramping needs when night falls that cannot be addressed easily without energy storage in the mix.

Batteries operate a lot like hydroelectric power. Hydro systems store potential energy in the water behind dams, and when the water is released, the potential energy converts to kinetic energy and flows through turbines to be converted to electricity.

Batteries are simple and have been around for hundreds of years. As an interesting and rather odd sidenote, Ben Franklin gave us the name for batteries, noting the similarity of a battery's stacked lead disks to the military fortifications of the time [2].

Batteries use the same principle of storing potential energy as hydroelectric systems, except instead of converting stored water potential energy to turbine-spinning kinetic energy, the battery uses a chemical process. Batteries have two electrical terminals, the anode and cathode, separated by a chemical material, the electrolyte (i.e., the chemistry of a battery). Electrons can move through the chemical circuit in both directions. Flow through the cathode provides a positive oxidizing effect on the electrolyte, storing energy. Flow through the anode provides the opposite effect, releasing the electrolyte's stored energy.

The CAISO duck curve phenomena was introduced in Chapter 2. Under the AB 2868 mandate, the California Public Utilities Commission (CPUC) incentivized investor-owned utilities (IOUs) to install energy storage systems 10 MW and greater at some feeder locations to address the duck curve phenomenon by providing voltage support and electrical islanding opportunities.

Energy storage systems provide an obvious solution to the duck curve that can store excess power to ensure grid thermal limits are not exceeded. Batteries can be charged during periods of low demand, low or negative energy costs, or overgeneration conditions. At dusk, energy ramping needs can be met by combining battery discharging and DR programs to eliminate firing peaker plants up or taking extreme measures such as rolling brownouts or blackouts.

Two Federal Energy Regulatory Commission (FERC) orders created new opportunities for utilities and VPPs operated by DER aggregators, introducing new business models that utilize grid-scale and virtual aggregated energy storage systems and allowing them to participate in a variety of grid support services and markets:

- *Order 841:* In 2018, FERC passed this landmark ruling, requiring the removal of energy storage barriers to entry in wholesale electricity markets.
- *Order 2222:* In 2020, FERC ordered wholesale market operators to allow DER aggregators to participate.

As mentioned in Chapter 2, if we consider neural grid structures as microgrid building blocks, energy storage is crucial to support the balance of supply and demand as well as power quality. As shown in Figure 5.2, battery energy storage deployments are growing rapidly as systems mature, become more interoperable, and become less expensive. Large commercial and industrial companies, microgrid and VPP vendors, and some homeowners are investing in energy storage systems to provide energy efficiency, sustainability, and resilience. According to EIA, the U.S. energy storage market will grow each year by 12 GWh by 2023.

5.5   Telecommunication Infrastructure

Reliable, low-latency communication capabilities are equally as important as reliable electron flow. Dependable communications are necessary for feedback and control. A combination of wired (copper and fiber) and wireless (radio, cellular, and microwave) make up the present communications framework.

5G networks are being added by telecom providers quickly and next generation wireless network services are being designed now. Each new generation of wireless networks brings greater capabilities that could have a significant impact on the electric power industry. 5G networks have theoretical latency values that are under a millisecond, with a fiber-like speed of 20 GB/s, 20 times faster than 4G LTE speeds of 1 GB/s. This amazing speed and low-latency capability make wireless technologies a very real alternative

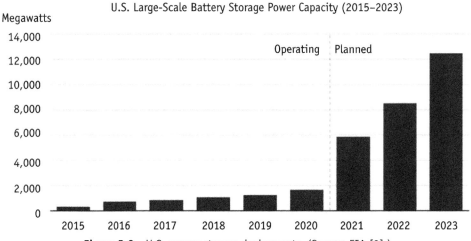

Figure 5.2   U.S. energy storage deployments. (Source: EIA [3].)

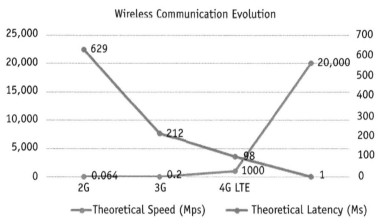

Figure 5.3 Wireless technology speed and latency.

to running fiber to every endpoint and can support large numbers of peer-to-peer equipment communication on the grid in front and behind the meter as well as grid-to-cloud communication. 5G is capable of establishing direct connections between grid assets and the tower, which provides minimal interference for those assets to communicate with one another or with the cloud in near real time. While 4G LTE can currently support most current and future grid communications, 5G is truly the technology with killer app potential for the energy industry. The energy IoT reference architecture will fully leverage these new capabilities to enable the neural grid described earlier.

5.6   Microgrids

In Chapter 2, we discussed the concept of thinking of everything as a microgrid—semi-independent building blocks to the overall grid. The author believes this is likely the end game for the electric power industry, but it is very unlikely to occur rapidly, even questionable to occur in our lifetimes.

   In 2012, the DoE defined a microgrid as: "a group of interconnected loads and distributed energy resources within clearly defined electrical boundaries that acts as a single controllable entity with respect to the grid. A microgrid can connect and disconnect from the grid to enable it to operate in both grid-connected and island mode" [4].

We think the DoE got it mostly right but are too restrictive when requiring a larger grid as part of the definition. There may not be a larger grid available for some microgrid examples. For instance, a remote island, a submarine, a ship, or a spacecraft is a microgrid. So is an electric vehicle. These examples of microgrids are not always (some are never) connected to a larger grid. However, for the purposes of this discussion, the DoE definition works very well.

Like DER and energy storage, microgrid technologies and deployments are rapidly advancing in capabilities, cost reductions, and popularity. Microgrids provide customers with resilience. Most microgrids operate in harmony with the grid when connected, but can operate independently when disconnected. Some island-based or remote microgrids routinely operate independently without the aid of a grid connection. Hence, the concept of "islanding" addresses the ability of a microgrid system to operate independently from the grid. The microgrid's ability to operate independently is the key aspect that provides resilience for the microgrid user. As catastrophic natural disasters increase with climate change events, and threats to grid operations increase due to man-made events such as breaches in cybersecurity, microgrid deployments rapidly become solutions providing energy surety (i.e., the U.S. Department of Defense (DoD) term for resilience) when the grid becomes unavailable for business owners, communities, and government agencies.

Moving beyond resilience, other compelling reasons for microgrid deployments include DER integration, carbon emissions reduction, power purchase risk reduction, and cost savings. Because of their smaller size and design to meet specific local requirements for energy, the microgrid generation asset mix more readily accommodates larger reliance on renewables and energy storage solutions, which supports cleaner power to help organizations meet their climate-friendly sustainability goals. Power purchase agreements (PPAs) can provide organizations with contractual electricity pricing over many years (typically 15 to 20 years), improving an organization's ability to accurately predict their electricity costs over time. Microgrid costs are now often lower than those provided by the local utility.

Many companies are cashing in on the trend through Energy as a Service solutions, where microgrid companies design, finance, build, and operate the system for their customers. Such companies include:

- Large multinational legacy electric power industry vendors: Schneider Electric, ABB, Hitachi, General Electric, and Siemens;

- Mid-level legacy and electric power industry vendors: Eaton, SEL, S&C Electric, and Emerson;

- Emerging electric power industry vendors and startups: Bloom Energy, Opus One, Spirae, CleanSpark, PowerSecure, AlphaStruxure, Enchanted Rock, Green Energy, and Scale Microgrid Solutions.

The microgrid EaaS market has become a very crowded space and there are simply too many microgrid EaaS companies to name them all. However, the trends are clear and microgrids are here to stay. It is also clear that many legacy and startup companies see enormous opportunities in providing EaaS through microgrids to some of the utility company's very best customers. Utility companies also recognize that customers seek resilience through microgrids, and many utilities have started deploying microgrids of their own, sometimes managing construction and operations themselves or outsourcing development and operations to third parties.

It is also noteworthy that utilities and EaaS companies can manage multi-DER systems such as microgrids and VPPs like they manage single DER systems. Figure 5.4 demonstrates how EaaS companies use energy IoT approaches to provide microgrid services to large numbers of customers in a mostly hands-off, low-maintenance, high-margin subscription business. Although a centralized command center or network operations center (NOC) may be needed, from an operations perspective, the service is primarily for supervisory oversight of a fleet of the operating company's microgrids. The service's primary mission is to step in only if there is an emergency and to ensure that the microgrids are performing to the contractual terms with the customer.

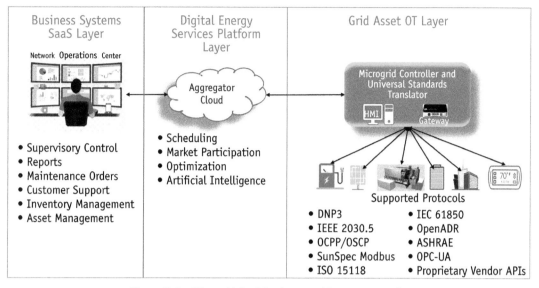

Figure 5.4 Microgrid EaaS business architecture example.

Microgrid EaaS companies can theoretically add new microgrids to their fleet simply and quickly by abstracting them so that they all look the same to the system, regardless of how the microgrid is configured at the local site. They create this abstraction through the cloud aggregation service and the local microgrid controller or gateway. The deployed microgrids are mostly autonomous at the local level using IoT technologies that leverage distributed (or edge) intelligence. To the operating company, a microgrid is a single point of aggregation that can be leveraged for a variety of different grid services and market participation. In these early days of microgrids, benefits for customers, vendors, and utilities will continue to be realized over the coming years as their capabilities grow over time and as costs continue to decrease.

5.7   Security

The electric power grid is a common target for bad actors who recognize that governments, industry, and people depend on it for basic needs. Grid disruption can be a disastrous event that causes monetary losses, industrial production losses, and even human life losses. It is also a very visible new topic that can get national and international attention. The grid is vulnerable to physical sabotage, equipment tampering, and energy theft and must be protected from physical and cybersecurity events. Sophisticated artificial intelligence analytics such as machine learning can be employed to monitor critical grid assets and communications traffic for anomalous or suspicious behavior. As these systems progress and learn, they will evolve to better protect the grid and strengthen security benefits for utilities and all stakeholders that depend on the grid.

Security will be discussed in more detail in Chapter 7.

5.8   Bulk Generation

Bulk generation remains a critical contributor to the overall generation needs to support today's electricity customers. Bulk generation is relatively inexpensive and can provide quality power to large populations. Problems associated with bulk generation include the overall cost of operations, the environmental impact of the fossil fuel power plants, and energy losses with long-distance transmission. In the coming years, bulk generation will almost certainly remain part of the grid's overall generation mix, but as more customer-owned DER assets enter the system, the amount of bulk generation needed could decline. Today's fossil-fueled bulk generation will be replaced by more economic and carbon policy-friendly grid-scale renewables and energy storage systems. Nuclear power remains a popular topic for bulk and smaller generation power needs, but sometimes take decades to permit and build and currently are quite expensive to build and operate.

The EIA reported that [5]: "Scheduled capacity retirements (11 GW) for 2020 will primarily be driven by coal (51%), followed by natural gas (33%) and nuclear (14%)." The majority of bulk generator capacity retirements were coal plants in Ohio and Kentucky, with some retirements of 1950s and 1960s era natural gas plants. EIA reported that 42 GW of planned additions for bulk generation would be made in 2020, with 76% (32 GW) intermittent renewable energy resources (wind and solar PV) [5]. The year 2020 saw the number and capacity of wind generation break the previous record of 13.2 GW (2012) by nearly 50%, 18.5 GW [5]. Figure 5.5 shows U.S. bulk generation additions and retirements for 2020, illustrat-

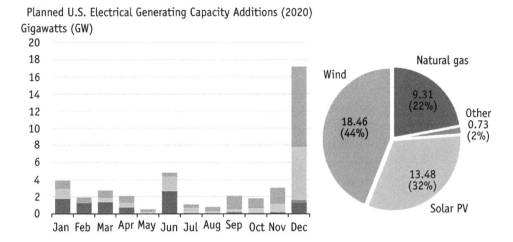

Planned U.S. Electrical Generating Capacity Additions (2020)
Gigawatts (GW)

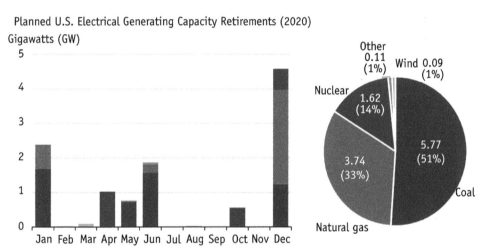

Planned U.S. Electrical Generating Capacity Retirements (2020)
Gigawatts (GW)

Figure 5.5 The 2020 planned bulk power plant retirements and additions. (Source: EIA.)

ing the transition from coal and natural gas to wind and solar PV systems.

5.9 Smart Homes, Buildings, and Cities

The race for home automation capabilities for the masses is finally here again. However, this time, affordably priced home automation consumer devices are IP network-enabled and can easily connect to high-speed home area networks. Major technology companies are peddling AI-enabled devices such as Amazon Alexa (Echo), Google Assistant (Home Hub), and Apple Siri (HomePod), and compatible smart devices such as smart plugs, thermostats, and surveillance cameras. These systems allow the user to set behaviors or policies that can be executed by the home automation devices and will easily and seamlessly be able to support energy consumption and generation devices autonomously to save money, provide resilience, and maximize the use of green energy. It will not be long before end-user customers will be able to leverage home automation devices to participate in utility customer programs or even in new local electricity market opportunities.

Imagine soon when a home energy device or collection of energy devices (i.e., thermostat and HVAC, rooftop solar, electric vehicle, battery, or something not yet imagined) can all be managed by an intelligent home energy management robot working in partnership with a much larger ecosystem that balances energy consumption with energy generation from the bottom up. Such a goal could also be accomplished with a smart gateway, smart meter, smart inverter, or another intelligent device.

Some commercially available systems exist that include energy management and universal protocol translator gateways inside the home or business. These systems can act as both the point of aggregation and also deliver continuous energy efficiency and carbon reduction improvements for the home or business owner. This single point of aggregated intelligence is responsible for performing all the duties of a home energy management system, including the efficient management of on-site generation, energy consumption, and storage. In time, these gateway devices may support participation in existing markets and possibly allow the home to operate as an islanded microgrid for some period of time if necessary. In addition, the gateway provides a standards-based coordination point that can allow utilities to use customer assets for grid services that support the utility's operations and grid optimization.

As witnessed with the personal computer and router industries in the 1990s, the plug-and-play experience that users have today started with interoperability standards. The same will be true in the energy industry as energy device manufacturers and technology companies work together to define features and interoperability. Numerous standards exist today and will be discussed throughout this book and in the appendix. Historically, interoperability across the electric power industry has been an enormous challenge, but standards such as IEEE 2030.5 for home area networks and OpenADR for load management are gaining traction to provide interoperability from centralized utility systems all the way to behind-the-meter customer assets. It took years for computer peripherals to become true plug-and-play devices, and a similar amount of time will be needed for smart electrical devices.

5.9.1 EVs

EVs are rapidly overtaking the automobile market as transportation electrification becomes technically achievable and better for the environment. They offer both a unique challenge to the electric power industry and new opportunities for energy suppliers, as electricity demand increases to meet the needs of the transportation industry.

Like batteries, EVs are another form of special DERs. Indeed, EVs can be thought of as mobile energy storage DERs that can show up at any time anywhere on the grid. The International Energy Agency (IEA) data illustrated in Figure 5.6 shows the steady increase in vehicle electrification from 2015 to 2020. The year 2021 showed even greater worldwide adoption. Data from the EVVolumes.com's Electric Vehicle World Sales Database showed that EV sales increased by 168% in 2021 versus 2020 worldwide [6]. Policy requirements and incentives to reduce pollution and GHG have led to greater EV adoption in China and Europe. China is the largest EV market in the world with 2.8 million EVs sold in 2021 [7]. In Norway, EV sales in 2021 were 65% of total vehicle sales [8], which makes them the country with the largest market share in the world [9].

Major car manufacturers around the world are announcing EV lines that are already here or will be available soon. The list includes such globally-recognized names as Audi, BMW, Ford, GM, Honda, Kia, Mazda, Nissan, Tesla, Toyota, and Volkswagen. Huge investments are being made by nearly all the automobile manufacturers to pivot from ICE vehicles to electric.

No one anticipated how quickly EV adoption would accelerate. Rapid growth in EVs is becoming a key challenge to utilities and the grid infrastructure, as large numbers of EVs spawn

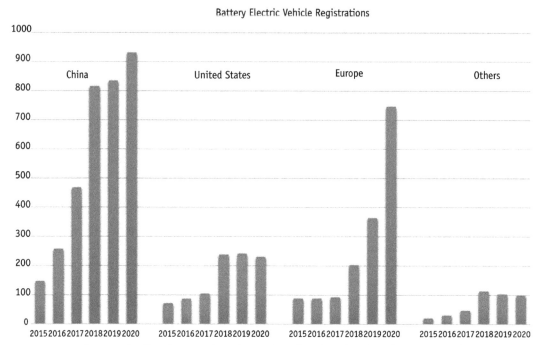

Figure 5.6   Electric car registrations and market share in selected countries and regions, 2015–2020, IEA, Paris [10].

enormous electricity demand that, if not coordinated, could push transformers, wires, and other grid assets to their thermal limits. The DoE predicts that annual electricity demand from a single EV will require 3.8 MWh of energy generation, requiring in aggregate as much as 26 TWh of annual incremental energy generation requirements by 2030 [11]. This is a staggering amount of load growth equivalent to the electricity demand of approximately 5.2 million households. Similar growth of this magnitude was seen in the 1970s when household air conditioning became commonplace.

Figure 5.7 shows two charts. The first chart from the EIA shows projected annual incremental generation growth requirements in low, medium, and high EV penetration scenarios. The second chart shows the cumulative projected generation growth requirements from 2021 to 2030. The second chart is based on data from EVAdoption.com and aligns closest with the EIA's Medium EV adoption projections. The charts show anticipated generation capacity growth requirements according to EIA and EVAdoption.

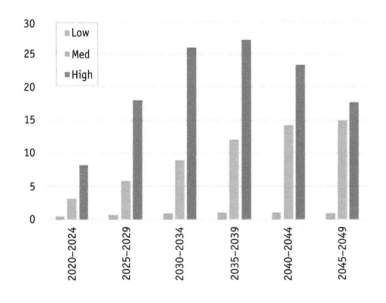

EV TWh Cumulative Capacity Needs

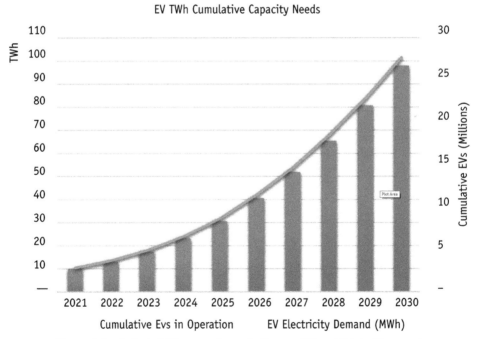

Figure 5.7 Projected EV energy demand. (Source: EIA and EVAdoption.com.)

com. The two charts show basic agreement in the amount of energy generation capacity growth that will soon be needed.

The United States currently generates approximately 4,000 TWh per year. Assuming a 28% EV market share by 2025 (4.72 million vehicles), an additional 102 TWh of capacity (2.5% increase) will be needed. If the auto industry meets the government challenge of supplying 50% EV sales (8.5 million vehicles) by 2030, that equates to an additional cumulative generation capacity growth of 2.9% or 116 TWh.

The electric power industry is currently experiencing one of its most transformational catalysts with the rapid adoption of EVs. EVs create new requirements for energy management systems to have better awareness for when they are plugged in and charging. They also introduce completely different protection system requirements and the ability to cohesively coordinate charge and discharge for large numbers of EVs at each grid constraint location serving them. Centralized command and control systems were not designed for situational awareness or coordination of EVs that randomly enter and exit the grid. More granular grid intelligence is required to support the coordination and scheduling of charging, discharging, and electricity storage of mobile DERs. Common microservices and message standards will help to coordinate EV load management and provide grid services that provide greater grid stability and protect the grid system from sudden or dangerous loads caused by these new EV challenges. Systems will need to evolve that recognize when an EV plugs in, and understand the local grid constraints, and charge point services can be scheduled that can meet the needs of the EV owner without compromising other grid assets.

5.10   EV Supply Equipment Charging Infrastructure

Although the EVs themselves have intelligence, the point of interoperability with electric power industry systems is the charging infrastructure. The EV can be described as an islanded microgrid when actively being used for transportation and only when it connects through the charging infrastructure does it become a component of the energy IoT ecosystem.

The emerging standard in this area is the Open Charge Point Protocol (OCPP) and the Open Smart Charging Protocol (OSCP), which are rapidly evolving standards being led by the Open

Charge Alliance. OCPP is the protocol used to support interoperability between the EV supply equipment (EVSE) and the charging point operator (CPO). OSCP is additional functionality for interoperability with the CPO and other entities, such as the utility, requiring capacity, and availability information. Figure 5.8 illustrates the IT/OT and energy IoT domains for EV charging along with the current leading standards for interoperability.

5.10.1   EV Fleet Charging

The most challenging issue facing utilities today is how to integrate and coordinate with EV fleets, where a concentration of charging EVs in a single location, each with large batteries, dramatically increases the historic site load. Light, medium, and heavy-duty EV fleets will primarily be located on distribution networks. Light and medium-duty vehicles will primarily charge at night because they will operate during daytime work hours. Heavy-duty vehicles will charge 24/7.

Figure 5.9 shows nominal maximum capacities for light, medium, and heavy-duty EVs based on data from Oak Ridge National Laboratory. When fleets of these vehicles are concentrated at a local grid access point, a new paradigm will be needed to ensure that charging requirements can be met without damaging the grid or causing grid instability. In most cases, charging required by EV

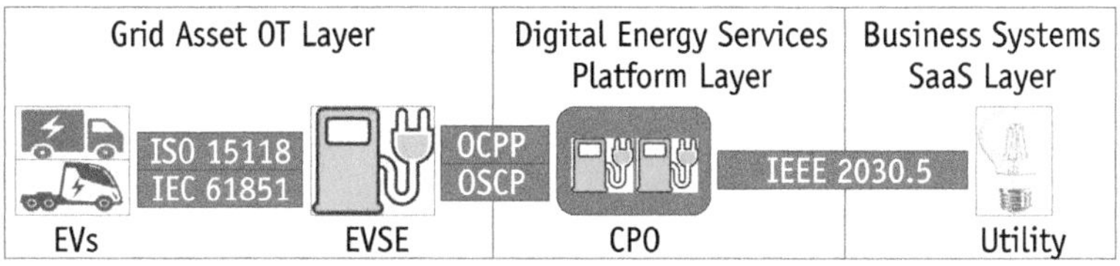

Figure 5.8   Standards view of EV-charging infrastructure.

EV Type	Max Capacity (KWh)
Light Duty	100
Medium Duty	350
Heavy Duty	1000

Figure 5.9   Nominal maximum EV capacity. (Source: ORNL [12].)

fleets will not be met solely by the grid. Transportation microgrids will need to provide some or most of the charging needs to ensure that other utility customers on the same feeder or distribution network do not experience adverse effects when grid constraints are exceeded.

Integrating transportation microgrids as part of EV fleet charging solutions introduces an interesting business opportunity for utilities, with the opportunity for the transportation microgrid to provide grid services for supply and demand balance, grid stability, reliability, and resilience. For instance, when medium-duty EV fleet vehicles are on the road during the day, the utility could use excess power generated by customer and grid-scale PV solar to charge transportation microgrid energy storage devices and create a business case for low (or negative) cost electricity during daylight hours. It could also use the transportation microgrid for ancillary grid services such as voltage regulation.

Figure 5.10 illustrates the potential opportunity for utilities with EV fleet transportation microgrids. Utility coordination of aggregated EV assets that are managed within grid constraints can enable grid and load-serving services for the utility in a mutually beneficial way for fleet managers.

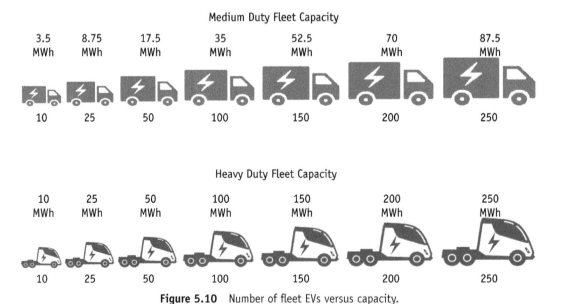

Figure 5.10   Number of fleet EVs versus capacity.

The aggregated EVs within the fleet can be coordinated much like a utility-scale battery when operated through a single control point such as that depicted in Figure 5.11.

How will this be accomplished? Refer to Figure 5.7. The standards exist today, and we can foresee IEEE 2030.5 solutions that can turn the problem of EV fleet coordination into an opportunity for utilities.

5.11   Conclusion

The present grid is an assortment of legacy assets that struggle to meet the needs of the populations they serve in most territories. The energy demands of the twenty-first century continue to increase worldwide, as well as stakeholder expectations for electricity reliability, power quality, transportation electrification, and scalability to support large numbers of renewables and other DER assets. More investment will be required to transition today's grid to one that is much nimbler and more adaptable. This will require innovation and careful planning aligned with significant infrastructure investment over time.

Unfortunately, the current utility systems of today that use legacy centralized command and control and one-way power flow, top-down architectures are unlikely to meet society's increased expectations for decarbonization and much more distributed assets,

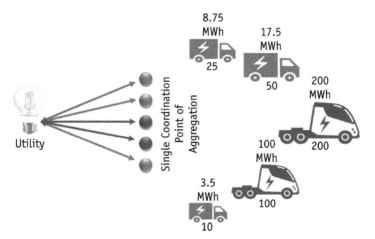

Figure 5.11   EV fleets provide utilities with grid-scale battery aggregation opportunity.

with most owned by customers and not the utility. The electric power industry must think differently. It must consider the same scaling issues and solutions that high transaction volume internet e-commerce companies like Amazon faced and conquered. We must leverage the same scalable platforms and advanced technologies supported by cloud providers. We must also prepare to leverage emerging and future technologies such as 5G communications, high-penetration energy storage, autonomous electric vehicles, and ubiquitous solar generation. We must look towards a future architecture that is event-driven, data-centric, bottom-up, and secure: one that can scale, supports many business models, and is inclusive. The time to begin this transition is now. IoT is the practical path forward.

References

[1] National Association of Regulatory Utility Commissioners (NARUC), https://www.naruc.org/cpi-1/energy-infrastructure-modernization/smart-grid/interoperability-glossary.

[2] UPS Battery Center.com, https://www.upsbatterycenter.com/blog/contribution-benjamin-franklin-electricity/.

[3] U.S. Department of Energy, Energy Information Agency, https://www.eia.gov/todayinenergy/detail.php?id=49236.

[4] Ton, D. T., and M. A. Smith, "The U.S. Department of Energy's Microgrid Initiative," *The Electricity Journal,* vol. 25, no. 8, 2012, http://dx.doi.org/10.1016/j.tej.2012.09.01.

[5] U.S. Energy Information Administration (EIA), Preliminary Monthly Electric Generator Inventory, https://www.eia.gov/todayinenergy/detail.php?id=42495, January 2020.

[6] EV Volumes, The Electric Vehicle World Sales Database, https://www.ev-volumes.com, January 2022.

[7] Bloomberg, https://www.bloomberg.com/news/articles/2021-12-08/china-new-energy-car-sales-spike-as-consumers-embrace-electric, January 2022.

[8] Yale E360, https://e360.yale.edu/digest/evs-made-up-two-thirds-of-new-cars-sales-in-norway-last-year, January 2022.

[9] Grid Integration Tech Team and Integrated Systems Analysis Tech Team Summary Report on EVs at Scale and the U.S. Electric Power System, https://www.energy.gov/sites/prod/files/2019/12/f69/GITT%20ISATT%20EVs%20at%20Scale%20Grid%20Summary%20Report%20FINAL%20Nov2019.pdf, November 2019.

[10]   IEA, Paris https://www.iea.org/data-andstatistics/charts/electric-car-registrations-and-market-share-in-selected-countries-andregions-2015-2020.

[11]   Grid Integration Tech Team and Integrated Systems Analysis Tech Team Summary Report on EVs at Scale and the U.S. Electric Power System, https://www.energy.gov/sites/prod/files/2019/12/f69/GITT%20ISATT%20EVs%20at%20Scale%20Grid%20Summary%20Report%20FINAL%20Nov2019.pdf, November 2019.

[12]   Smith, D., R. Graves, B. Ozpineci, and P. T. Jones, "Medium- and Heavy-Duty Vehicle Electrification—An Assessment of Technology and Knowledge Gaps," Oak Ridge National Laboratory, December 2019, https://info.ornl.gov/sites/publications/Files/Pub136575.pdf, Oak Ridge.

6

Energy Business Systems (SaaS) Domain

This chapter focuses on the virtual energy systems SaaS domain within the energy IoT reference architecture, which is highlighted in Figure 6.1 by the red outlined cloud on the left side. The energy systems SaaS domain that we envision:

- Provides end-to-end cybersecurity, only providing access to authorized and authenticated users and systems, and using Transport Layer Security (TLS) protocols for encryption.
- Uses abstraction to connect to grid assets through the digital energy services platform (Green Cloud).
- Provides services that are interoperable with any authenticated and authorized subscribers, preventing data and functionality silos and leverages message bus communications such as publication/subscribe (pub/sub).
- Utilizes industry protocol and semantic information model standards to support more interoperability.
- Leverages the Green Cloud services layer to store data and reduce integration time and costs.
- Provides APIs and clearly defined interfaces to support innovation and integration with existing and yet-to-be-invented new systems.

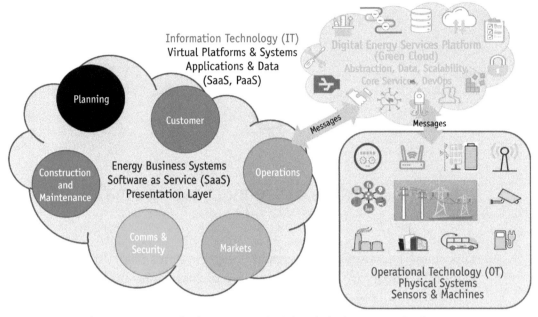

Figure 6.1 Energy business systems SaaS domain in the energy IoT reference architecture.

One of the core principles within the energy IoT reference architecture is that the energy business SaaS system does not connect directly to physical OT devices for two very good reasons.

1. When using a direct connection to a physical grid asset, if the capabilities of an asset change for any reason (new feature, bug fix, new standard support, change in a standard, or security change), every device and its connection to system components also need to be changed, tested, and reprovisioned. Conversely, within the energy IoT reference architecture, the Green Cloud utilizes an abstraction layer using virtual machines and containers (Docker) and orchestration (Kubernetes, pronounced koo-ber-net-eez) combined with digital twin agents[1] to push new capabilities, bug fixes, and upgrades universally to edge devices. Using these technologies solves problems with managing

---

1. The concept of digital twin agents has been discussed internally by the author and past contributors at great length. Agreement is not unanimous, but consensus exists. We will introduce the author's concept of the digital twin agent with much greater detail in Chapter 7.

version control and simplifies deployment, as the software can be changed one time and will immediately propagate to any system using that newly modified asset type.

2. Further, by leveraging the Green Cloud abstraction layer, the data is decoupled from the business systems to create much richer chances for analytics and interoperability between other systems and services. This decoupling of data from systems is a key challenge to utility systems today and is one of the primary issues identified by the DoE's PNNL in its grid architecture theory. PNNL's grid architecture is discussed in more detail in Chapter 10.

6.1   Planning Systems

Planning groups and the forecasting systems that they use provide utilities, system operators, and third parties with algorithmic predictions of how the grid network will perform going forward and what the grid will need in order to support future loads and operate within industry tolerances.

6.1.1   Long-Term Planning (LTP)

LTP, typically found within the utility's distribution planning group, performs planning on a time horizon of greater than 30 days. The distribution planning group performs performance analysis to plan the future distribution infrastructure needs of the system. This analysis includes load and distributed generation forecasts using spreadsheets and sophisticated tools to balance the system and identify grid upgrades to support performance metrics. More recent LTP analyses include capacity planning and nonwire alternatives (NWAs) that consider implementation of DER (e.g., energy storage, microgrids, solar) instead of traditional infrastructure construction projects or upgrades such as "re-conductoring."

6.1.2　Short-Term Planning (STP)

STP groups work on much shorter time horizons than LTP, measured in hours, days, and weeks. The STP group uses real-time system data, recent historical data, and other variables such as weather and market price forecasting to determine short-term grid needs. The STP group works closely with the distribution grid operator (DGO) and market operators to provide hour-ahead, day-ahead, week-ahead, and sometimes shorter-term forecasting to balance load and generation and to ensure the grid operates within tolerances. The introduction and higher penetration of renewables and other forms of distributed generation are being factored into the calculations. This role is a critical part of grid operations and is necessary to manage and adapt to near real time and short-term coordination of assets to optimize operational and financial performance for the utility.

6.1.3　Weather and Load Forecasting

Weather and load forecasting, mandatory with both LTP and STP, forecasts both short-term and long-term grid behavior and needs. Obviously, weather-dependent solar PV and wind generation output profiles are directly affected by the amount of solar radiation or the wind speed and direction. Weather satellites and large databases of historical weather information provide forecasting systems with valuable and much more granular data than in the past. Weather forecasting systems are evolving, using historical data and modern algorithmic analysis capabilities such as machine learning that provide more granular analytics and better forecasting accuracy. New grid assets include weather metrology devices that can provide granular weather data at very reasonable prices.

6.2　Customer Systems

Utilities and third-party energy providers use customer systems to manage their relationships with the end user. Customer information includes names, organizations, geospatial, roles, types of services provided, and other relevant information for each individual customer. This may include information on the customer's load history, energy assets, point of interconnect, and possibly even relationships with other providers or market participation.

6.2.1 Customer Programs

Utilities provide voluntary customer programs for customers to participate in helping the utility with energy efficiency optimization, managing peak load conditions, and complying with policy mandates. Utilities reward participants through new rate designs, reduced energy costs, and other financial incentives to change their energy usage behavior. Demand-side management customer programs have been around for years to manage peak load and energy efficiency. In addition to DR programs, utilities are now also implementing supply-side customer programs that leverage customer-owned distributed generation to provide load balancing and ancillary services. However, demand-side management customer programs are much more common and can leverage common customer load assets such as water heaters, thermostats, and pool pumps. Unfortunately, one of the most popular DR customer programs, thermostat control, can yield unreliable demand reduction results because customers can elect not to participate or can simply change the thermostat setting prior to a DR event and still achieve the program reward without truly reducing demand when it was needed most.

Recently, the idea of local markets has become a popular topic that could reward customers that own distributed generation or flexible load assets. If a DSO was to support customers with demand side local markets on a local distribution network, the concept of bidding 4°F into a DR market would have no meaning. Instead, a home energy management system or similar application would, for example, bid 2 KWh of load reduction from 5 p.m. to 6 p.m. to lower and measure power consumption using meaningful market values. In a distribution market where bids are KW, KWh, or another ancillary services market metric, some customer programs such as the thermostat example may not be useful. Over time, thermostat customer programs will likely be phased out and replaced by other aggregated capacity grid services using normal energy market pricing metrics and more open markets to homes, businesses, and third-party participants.

6.2.2   Metering Systems

Today's metering technologies are often called the "cash register" for the utility. AMI systems normally use a polling methodology over mesh networks to gather metrology data. AMI systems provide enormous value to utilities, not only providing usage data (KW and KWh) for billing purposes, but they can also provide last-gasp data to outage management systems (OMS) to accurately identify where outages occur and who is impacted. Other data such as power factor, voltage, current, and reverse power flows can be provided by smart meters. This data is analyzed for quality and collected at head-end systems. Energy usage data is used by CIS to create bills and present them to customers. It may also be used to validate that the customer has met its contractual obligations when participating in markets or utility customer programs via measurement and verification (M&V) systems.

AMI systems were discussed in Chapter 2 as one of the data silos that could use some rethinking. AMI systems can be quite expensive to implement and interoperability with other systems can be challenging. Data latency can also be a problem when timely data is needed by other systems. In Chapter 7, we introduce the concept of a digital twin agent. For metering solutions, the digital twin agent could replace the polling technologies commonly used in today's AMI systems and perform self-reporting at some interval (5, 10, 15, 30, 60 minutes), when something changes, or when an event occurs such as a DR program. The smart meter digital twin agent could provide accurate, secure, and timely data for authorized operational, M&V, market, customer billing, and other analytic systems.

6.2.3   Interconnect System

The plug-and-play grid of the future will require much more automated and simple interconnection agreement solutions than those found today. Interconnecting in front of and behind-the-meter DER assets can take several days or even weeks in the most efficient and sophisticated interconnect systems. Manual interconnect processes for larger DER or microgrids may require utility studies and can take months or even years to approve, sometimes resulting in the customer simply giving up and abandoning the process.

Identification and recognition of installed DER must become more automated, which provides the utility with the type of DER installed, its capacity, geospatial data, and its capabilities. Without this information, utility connectivity and power flow models are incomplete and will provide equally incomplete and erroneous grid state estimates and forecasts. The transformation to a highly efficient, carbon-friendly, DER-rich, plug-and-play power system will require changes to interconnect processes and systems and the DERs themselves. DER vendors will need to manufacture assets that self-announce, describe and provision themselves, and can quickly and automatically be available for integration with larger grid and market systems. Standards such as IEEE 2030.5 can help to support more automated interconnect processes and systems. This standard, and its DER discovery capabilities, will be discussed in Chapter 8.

6.2.4   M&V System

In most cases, the interaction between utility and customer occurs just once each month when the power bill is created and presented to the customer. Today's M&V systems are based on meter reads primarily for energy usage billing purposes. As more customer DER enters the transformed grid, M&V systems will evolve to support more utility customer programs and perhaps even local market participation in near real time. For example, market participants will bid energy, KWh, whether consumption, export, or another measurable ancillary service, to support distribution grid operations. The distribution market operator (DMO) M&V system will determine if the customer has met the contractual obligation, as anticipated during the time period defined in the market participant's bid. The market participant should have its own M&V system that includes metrology devices that can be used to perform validation. It is likely that the Green Cloud will have M&V services that can instantly determine whether the DMO and customer systems agree and flag the ones that do not for investigation. With these services, vendor companies will have new opportunities to provide building measurement hardware and cloud settlement services that significantly simplify and ensure fairness when open, democratized markets are enabled.

6.2.5 Customer Information (CIS), Settlement, and Billing Systems

CIS, settlement, and billing systems that track customer information like names, addresses, contact information, and customer program enrollments have been around for a long time. Settlement systems bill each customer based on M&V data, and billing systems arrange customer bills and collect payments. Existing CIS, settlement, and billing systems will almost certainly evolve, but existing systems are adequate to support an energy IoT transformation.

6.3 Operations Systems

Grid operations systems are the heartbeat of the transmission system operator (TSO) and DGO used to manage the real-time operations of the transmission and distribution grids. Operations systems address:

- *Grid situational awareness:* Systems that provide real-time visibility and monitoring of grid assets and overall status;
- *Command and control:* Systems that coordinate and manage different grid assets to maintain the transmission and distribution networks within acceptable operational tolerances.

Today's centralized SCADA systems used to communicate with grid assets are much different than those envisioned in the energy IoT reference architecture. Rather than register-based telemetry SCADA control, the energy IoT reference architecture envisions the use of standardized communication protocols and common semantic message payloads, pub/sub message buses, and digital twin agents. The result will be a nimbler and more adaptive grid that integrates new grid assets quickly and supports more granular, autonomous, resilient, and efficient grid operations. This will be discussed in more detail in Chapter 7.

6.3.1 Transmission Systems

Transmission systems provide grid operators with the tools to plan, coordinate, and manage high-voltage transmission networks. TSOs generate plans to meet capacity needs. They qualify bulk generation and load providers to ensure they have the financial and operational resources to meet contractual obligations when participating in bulk energy markets. They manage transmission operations to ensure that loads can be served and provide quality power to match those loads through bulk generation resources. Transmission EMS are equivalent systems to distribution management systems (DMS) used by distribution operators. EMS provide direct or indirect dispatch of bulk generation and load assets (such as energy consumption from grid-scale energy storage systems) to balance capacity needs. Transmission systems were designed with a top-down, one-way power concept. In the bottom-up future of an energy IoT architectural transformation, ISOs are envisioned to provide the same bulk power management role. However, with much more granular distribution-level information, that data will be shared in aggregated, rolled-up granular portions to provide better accuracy for the amount of bulk power required and where it is needed. Obviously, in a high-penetration DER world, the mix, location, and amount of generation will be primarily located on distribution networks and out of the control of the transmission operator. However, we will likely see more bulk generation DER such as grid-scale wind, solar, and energy storage. Fossil fuel bulk generation will gradually reduce, and the goal of carbon-friendly electricity supply will be achieved over time.

6.3.2   DGO Systems

DGO systems are managed by the DGO and provide them with the situational awareness and command and control capabilities necessary for safe and reliable distribution grid operations. Within the utility, this function is performed by the electric distribution operations group. SCADA, DMS, advanced DMS (ADMS), and OMS are the systems DGOs use to monitor grid performance and provide control signals to assets to maintain balance and to operate within tolerances. These systems can be very expensive, approaching hundreds of millions of dollars, and demand much care and feeding. Most distribution operations systems require customization and even minor grid infrastructure or system changes may require challenging system integration efforts.

Further complicating the DGO role, legacy SCADA telemetry systems are used to provide direct communication connections between grid assets and operational systems. Not only is this point-to-point communication mechanism difficult to change and complex to configure, it also requires highly specialized and often overworked SCADA engineers to generate a point within the SCADA system. SCADA software must translate that meaningless point to engineering units that can be digested by other systems. Troubleshooting these systems is complex with a variety of potential failure points. Problems may occur due to a mapping configuration mistake, poor or lost connectivity, incomplete or erroneous integration, data integrity or the physical asset itself, which could be a hardware malfunction, firmware issues, or something else. SCADA systems will need to evolve to support semantic language messaging payloads in the energy IoT reference architecture, or they will most likely be replaced by modern message buses and operational digital twin agent solutions.

6.3.3   DERMS

DERMS provide a mechanism for communicating with individual DER assets or through aggregated DERs provided by the utility or a third-party aggregator. Many people in the industry believe that current DERMS will become a piece of EMS, DMS, and ADMS and will no longer be separate systems. There has been a visible industry trend where large utility vendors are purchasing the smaller startup DERMS companies and integrating their solution into their own EMS, DMS, and ADMS.

Undoubtedly, the integration and coordination of DERs in a way that results in greater rather than poorer grid reliability are one of the biggest challenges currently facing utilities. Even if the utility knows that a particular DER asset is deployed (which most do not), utilities can only directly control registered DERs that are in front of the meter. This is not necessarily the fault of the utilities or the DERMS providers. Regulatory rules govern the reach of the utility, stopping utility control at the meter. This can be overcome by utilities by creating customer programs that incentivize the customer to provide some level of control of the behind-the-meter customer assets. Regulatory rules in other parts of the world may also allow utilities to operate customer-owned, behind-the-meter DER. However, utilities' use of centralized, point-to-point operations also introduces scale issues when coordinating with millions of customer-owned DERs. Also, nonhomogenous communications systems include integration across utility telemetry systems, public internet, and cellular networks, which can complicate reliable communications from the utility to the DER. Most of all, the use of protocols such as OPC-UA, DNP3, and Modbus require register mapping that can be time-consuming and prone to errors. The legacy register-based standards will likely be replaced over time by rich semantic information models that utilize pub/sub protocols (instead of point-to-point) with common message functional payload formats. This will result in better interoperability between utility systems and customer assets, but it will also result in lower integration time and costs and create smarter local systems where the assets can communicate with each other instead of a central operations center. This will allow smaller grid structures to manage themselves while also supporting larger structures. Existing information model standards such as IEC 61968/70 (CIM), IEC 61850, OpenFMB, OpenADR, and IEEE 2030.5 are likely candidates for this transition to more modern messaging payload approaches. In the short term, these modern standards can be supported by protocol adapters to speak the language of the legacy standards. This is a fundamental principle of the energy IoT reference architecture and one of the primary reasons that the Green Cloud is a radical yet practical need for the electric power industry going forward.

6.4 Market Systems

The legacy electricity wholesale market system has enabled utilities, independent power producers (IPPs), and now qualified third-party aggregators to trade and potentially generate profits in bulk energy, capacity, and ancillary services markets. As currently designed, the bulk market systems present smaller players with significant barriers to participation in these markets.

DER systems can now be purchased by businesses and homeowners at prices that yield justifiable returns on investment (ROI), with the best ROIs in areas with rising energy costs and falling DER prices. Regulators, ISOs, and utilities are evaluating the creation of new DSOs to animate new distribution markets and engage smaller DER owners. Imagine the impact of a transparent, open, and broadly inclusive market that accelerated the adoption of DER, created opportunities for innovation and new businesses, and completely changed the way that the grid operates and the way that consumers, producers, and machines interacted with it. As conversations about DSOs increase in number and intensity, distribution markets can be expected to emerge in the coming years, driving faster DER adoption as even individual homeowners can participate in grid operations.

6.4.1   Wholesale/Transmission Markets

Generally there are three types of transmission markets:

1. *Energy markets:* These markets are used by electricity suppliers to sell electricity to load-serving entities to meet demand. This type of market is an auction that clears when electricity supply and demand are in balance. Prices are based on megawatt hours (MWh) with day-ahead and real-time (or spot) markets as the norm.

2. *Capacity markets:* Also called reliability markets, capacity markets are used to provide additional reserve margin in case electricity supply and demand do not match for whatever reason. An example of an energy supplier might be a fossil fuel peaker plant, but there is also the possibility of a supplier that could reduce demand when needed.

3. *Ancillary services:* Ancillary services markets are services other than electricity supply and demand. These services are primarily to maintain proper frequencies and quick-response backup power if another generating unit fails; necessary services such as reactive power and voltage control that guarantees power quality and supply/demand imbalance thresholds are within tolerance.

   Electricity market instruments, processes, and systems are not uniform and vary depending on the country, region, state, or province governing them. Transmission market contracts can be bilateral agreements or auctions. Typically, day-ahead, hour-ahead, and real-time spot markets are supported, but there may be longer-term markets supported in some areas. Some bulk market operators provide local nodal markets and calculate locational marginal pricing for delivering power to a node at a specific location.

Transmission markets are always evolving, and the introduction of aggregated DER presents new opportunities for democratizing participation to smaller players. As the energy IoT reference architecture transformation evolves, the transmission markets will remain intact, but will likely introduce new data exchange and other technical requirements and policy changes for participation. The United States is experiencing some of these policy change effects as FERC Order 2222 mandates the inclusion of DER aggregators by transmission market operators.

6.4.2   Retail/Distribution Markets

In order to create a true new energy economy, organizational changes will be needed in addition to the changes needed architecturally, technically, and in bulk energy markets. The concept of local distribution energy markets will require new organizations to operate those markets. A DSO is one way that this might be accomplished. Local DSO-managed markets would catalyze a new ecosystem that empowers the utility to either operate the markets themselves as the DSO or to participate in local markets operated by a third-party DSO. These markets could serve utilities, prosumers, IPPs, third-party DER aggregators, and technology providers to create new business models that can profit and participate in providing grid services at the local level. This will result in cleaner and less expensive power, greater efficiencies, more reliable power, and greater resilience to grid disruptions.

There are several types of DSO models that cover a spectrum of different organizational styles. This spectrum, shown in Figure 6.2, ranges from highly decentralized peer-to-peer models to highly centralized transmission-level models. In between those two extremes are nodal models representing interaction between nodes, physical locations on the distribution network, such as a transmission-distribution (T-D) substation or a feeder. Organizing into nodes enables a separate nodal market instance for the actors on each local distribution area (LDA) node.

Regardless of which DSO organizational model is selected in any particular local region, the energy IoT reference architecture anticipates market services that will support a wide variety of

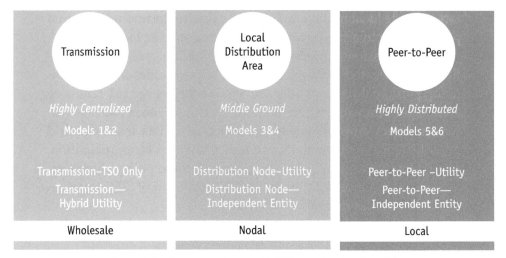

Figure 6.2   Spectrum of potential DSO organizational models.

retail market types. These local DSO market services will support the calculation of pricing, bids, and settlement. At present, these services do not exist. Transactive energy instruments will need to be developed by innovative companies to enable transactions between participants in new local markets.

6.4.3 Carbon Markets

Carbon markets are popular conversations with policymakers with the idea of taxing companies for their carbon emission contributions. According to the EIA, the United States generated 1.55 billion metric tons of carbon dioxide (CO_2) with 4.01 trillion KWh generated by the electric power industry in 2020 (or about 0.85 pounds of CO_2 per KWh) [1].

Surprisingly, even some traditional oil and gas companies like British Petroleum (BP) publicly advocate for carbon market instruments. The Climate Leadership Council, a bipartisan committee of U.S. Senators, recommended a $40/ton tax for companies. They further recommended that proceeds would be distributed to Americans on a quarterly basis. Bold discussions around carbon markets are ongoing at the local and national levels. In France, all businesses must publicly post their annual CO_2 emissions. It is probably a matter of time before these markets and the measurement of carbon emissions become a reality in several parts of the world, perhaps even the United States.

6.5   Communications and Security

The evolution to 5G networks will have a profound and positive impact on utility operations. Early deployments of 5G are already being seen in many parts of the world and we are even seeing 3G technologies being completely phased out. 5G is coming sooner than one might think, and grid systems should be designed for much faster (up to 20 times over 4G LTE) fiber-like speeds and always-connected IoT devices. However, system designs must account for expected communication losses with the ability to continue operations autonomously in the last authorized command set received or a default operational mode.

6.5.1   Cybersecurity

Cybersecurity systems include private/public key management, identity management, and role-based access management, all of which are a major concern for grid operators. These systems support message encryption, ensuring that only authorized and authenticated actors have access to systems, services, and assets.

6.5.2   Network and Telecom Management

Network and telecom management systems provide configuration and health monitoring capabilities for operational and market needs, which may include predictive analytics and other analytics that search for tampering, sabotage, or other bad actor activities that target grid telecom networks.

6.5.3   Physical Security

Physical security systems, equally important as cybersecurity, protect physical assets from bad actors. These systems could include public safety applications that perform facial or other biometric recognition to flag potential intruders or to automatically provide access to restricted areas to authorized users. Other physical security systems such as gunshot detection are becoming more prominent around substations and other critical grid assets.

6.6   Construction and Maintenance

The grid, the electric power industry's infrastructure, is the largest interconnected machine in the world, with the U.S. grid alone comprising over 360,000 miles of transmission lines and 6,300,000 miles of distribution lines [2]. Aging grid infrastructure often remains in service well beyond the expected service life. It needs to be maintained and replaced by new infrastructure using new technologies that are smarter and cleaner and can scale to meet the growing power needs of customers. New grid smart infrastructure should be designed to be less costly to maintain and be extended over time to meet future changes to operational and communications systems.

6.6.1   Asset Management Systems

Utilities and electricity service providers must maintain a repository of the assets that it owns and maintains. Asset management systems are the systems used to recognize the devices that they have deployed in the field. They provide many different functions including managing the asset's life cycle, spare part inventories, and scheduled maintenance. The asset's life-cycle information is collected on everything from design, installation and commissioning, operational performance, maintenance history, system upgrades, replacement, and decommissioning. Asset management systems are a critical and necessary solution to manage risk, maximize asset performance, and optimize asset and system costs. The system can interoperate with workforce management systems to create maintenance and repair tickets and optimize the available maintenance teams based on their location and skills.

6.6.2   Workforce Management Systems

Workforce management systems are designed to create tickets for workers to perform routine and unscheduled maintenance efforts. They create tickets to prioritize maintenance and replacement of grid assets. Modern solutions recognize the location and skillset of workers to optimally route the correct resources to perform productive work in the minimum amount of time. Workforce management systems have become quite sophisticated, leveraging new analytic technologies and mobile solutions that can track progress and create real-time communications with central work dispatch operators and the workers in the field.

6.6.3 Geospatial Information System (GIS)

Another critical and necessary utility system is the GIS. With the full deployment of the Global Positioning System (GPS) in 1993 [3], the ability to precisely locate anything on Earth launched the GIS industry. Current GPS receivers provide accuracy within 5.97 ft (1.82m) 95% of the time [4]. These systems provide useful asset locational information that supports utility connectivity and power flow models that provide utilities with state estimation and forecasting capabilities. They also support workflow management systems when dispatching maintenance crews, routing them to the correct location during construction and maintenance events.

Asset connectivity information is provided by GIS to be used by systems that simulate power flow to provide current state information and forecast future behavior. Connectivity models provide the power flow model with the available asset locations on the grid network. The power flow model, in turn, is used to predict grid needs over some time period (real time, minutes/hours ahead, days/weeks ahead). As we move towards a neural grid that is more dynamic, nimble, and adaptable, the GIS is required to sustain models to support grid operations teams and market participation. Precise locations of grid assets is critical. Within the energy IoT energy system domain, all seven of the subdomains (planning, customer, operations, markets, communications and security, construction, and maintenance) require interaction with the GIS.

6.7 Conclusion

One of the primary challenges with today's integration, scaling, and data silo issues is a result of how systems within the energy systems domain were architected and developed. Integrating DER assets into today's architecture is complicated and time-consuming. Integration with other energy system solutions often requires "spaghetti" custom interfaces that need to be maintained and tested when other code changes are made. The unfortunate reality is that many of the energy systems were designed as siloed systems. Existing systems collect data directly from grid assets, store the data in proprietary data repositories, and have latency issues when sharing data with other systems.

Some of the most profitable and successful technology companies learned valuable lessons about their systems and the data collected by them. What they learned is that the systems are not the most important part of the architecture. In fact, it is all about the data. The energy IoT reference architecture does not even use system names such as DMS, SCADA, AMI, AMS, Workforce Management System (WMS), or GIS. Rather, system functions drive the discussion around functionality and data requirements, avoiding the urge to solve industry needs with legacy systems. As an industry, investment decisions often begin with conversations about which legacy systems are needed, forcing decisions based on existing architecture rather than ecosystem functional requirements.

Shifting the conversation to true data requirements is a vital step. Contrast industry conversations that start with "We need an AMI system," instead of "We need to collect and store energy usage data for M&V, settlement, billing, and analysis," or "We require a GIS," rather than "We require precise geospatial locational data for every asset for power flow modeling, operations, and planning." Starting investment discussions with data requirements leads to scalable, message-based, interoperable systems that are easier to maintain and evolve, cost less to implement, and create real stakeholder value and opportunity.

A contemporary, data-centric architectural construct must be introduced. Chapter 7 will describe the centerpiece of the energy IoT architecture, the digital energy services platform (Green Cloud) domain. This new architectural layer democratizes data, provides services for systems and OT assets to abstract and simplify communications, and includes a DevOps environment to create energy-specific applications, services, systems, and solutions.

References

[1]   U.S. Energy Information Administration (EIA), "Frequently Asked Questions: How Much Carbon Dioxide Is Produced per Kilowatthour of U.S. Electricity Generation?" https://www.eia.gov/tools/faqs/faq.php?id=74&t=11#:~:text=In%202020%2C%20total%20U.S.%20electricity,CO2%20emissions%20per%20kWh., 2021.

[2]   Office of Electricity Delivery and Energy Reliability, U.S. Department of Energy, "United States Electricity Industry Primer," July 2015.

[3]   NASA, "Global Positioning System History," https://www.nasa.gov/directorates/heo/scan/communications/policy/GPS_History.html, August 2017.

[4]   GPS.gov, "GPS Accuracy," https://www.gps.gov/systems/gps/performance/accuracy/.

7

Digital Energy Platform Services Domain: The Green Cloud

The centerpiece of the energy IoT reference architecture is the digital energy service platform or Green Cloud, which provides the elastic scalability and abstraction necessary to simplify communications between SaaS applications and edge devices. This chapter is focused on explaining the Green Cloud's function and its architectural components within the energy IoT reference architecture.

7.1 Architectural Principles of the Green Cloud

The Green Cloud is the heart of the energy IoT reference architecture. Five core architectural principles are supported throughout the Green Cloud domain: scalability, abstraction and interoperability, reduced complexity, loose coupling, and business model innovation.

7.1.1 The Principle of Scalability

Scalability is the ability to scale in order to support ever growing numbers of energy assets; this is undoubtedly the biggest challenge that the electric power industry faces today. The Green Cloud platform includes standard cloud vendor scaling tools. Large Cloud vendors such as Amazon Web Services (AWS), Microsoft Azure, Google Cloud, and others provide this out-of-the-box capability for elastic scaling that can spin virtual machines up and down to match computing needs. They also include management solutions for big data capable of handling the massive amount of data that electric power systems and devices can generate.

7.1.2   The Principle of Abstraction and Interoperability

The lack of interoperability between systems and devices is undoubtably a consultant's best dream. Without interoperability, hundreds or thousands of consulting hours may be needed to create point-to-point interfaces or APIs to integrate one system with another. Beyond the cost in time and dollars, this approach can create very brittle systems with unintended and unexpected consequences when one part of a system changes and breaks other parts of the system. Further, these rickety systems become maintenance challenges as new devices or new capabilities are added requiring extensive regression testing.

Even if one edge device provides the same functionality as another, they will still differ in other ways. Registry values and locations, for instance, will almost certainly differ between manufacturers. The capabilities of one manufacturer will not be supported by another. Fortunately, adapters can make similar edge devices look the same and the Green Cloud is the logical place where these adapters can abstract how SaaS systems communicate with edge devices, regardless of who built them.

Abstraction and interoperability are key architectural principles of the energy IoT reference architecture with the ultimate goal for plug-and-play interoperability. Abstraction simplifies communication between the Cloud and edge and can be accomplished by utilizing a common messaging approach and an intelligent edge gateway that can take those messages and parse them into asset-specific languages (e.g., Modbus, DNP3, OPC-UA). IEEE 2030.5 shows promise of being that standardized abstraction messaging approach and is rapidly being adopted by energy vendors. Interoperability is definitely a challenge in the electric power industry with large numbers of assets, vendors, and protocols in use. Today's interoperability is achieved through adapters that translate one protocol to another. Message-based protocols such as

IEEE 2030.5 again show promise for simplifying interoperability, and mandates such as Rule 21 in California are designed to accelerate the use of message-based communications all the way to end devices. In California, IEEE 2030.5 is the default standard that vendors must support for grid-connected assets. As vendors move towards this approach, the panacea of true plug-and-play interoperability may be achieved with the ability to discover assets as they connect or disconnect from the system.

The three-layer IoT architectural approach addresses interoperability by using the Green Cloud platform as the interoperation layer. The platform includes numerous tools to simplify and speed integration efforts. Again, the ultimate end game is a true plug-and-play ecosystem that provides integrators with standards-driven microservices and adapters for integration, digital twins, big data scalability, and trust services.

7.1.3   The Principle of Reduced Complexity

The electric power industry's historic strategy of using the same engineering solutions of the past to manage the rising complexity of today is reaching its fundamental limits. An IoT architectural approach is the practical mechanism to reduce complexity and overcome limits on the number of connected devices and the amount of data that can be processed. Today's highly complex centralized systems concentrate intelligence within the enterprise by using applications to "think" and make decisions for edge assets. However, these systems become less effective as the number of edge devices grows and the complexity becomes overwhelming with hundreds of thousands of edge devices needing coordination. The architectural principle of reduced complexity recognizes that the intelligence in edge devices can be harnessed to coordinate with other DER and larger hierarchical grid structures. Intelligent edge devices have the ability to rapidly communicate with one another and make local decisions to manage energy. Round-trip communication between edge devices and SaaS enterprise solutions is becoming less of a requirement, reducing the opportunity for latency and enormous computing requirements at central operations.

Communications between massive numbers of DERs is becoming a fundamental requirement to provide necessary reliability and resilience for the grid. Common information models, asset grouping or aggregation, scheduling systems, and messaging protocols provide the technology needed to coordinate with large numbers of intelligent edge devices. Standards such as IEEE 2030.5 and OpenFMB were designed to support peer-to-peer asset communication and coordinate among individual DERs and groups that take advantage of edge device intelligence.

7.1.4   The Principle of Loose Coupling

Loosely coupled architectures are component-based. Applications and systems can be stitched together using software components and additional logic to support a business or technology use case. Ideally, the components are reusable and capable of supporting multiple applications and use cases.

The Green Cloud provides reusable message-based microservices that allow SaaS and edge systems to leverage with minimum or no dependencies to other microservices. Containers and orchestration use loose coupling to push minimum footprint capabilities to edge devices. SaaS applications leverage loosely coupled microservices to decrease development time, automatically upgrade capabilities and fix bugs with upgrades to underlying microservices.

7.1.5   The Principle of Business Model Innovation

As technology and business model innovation engage in a virtuous cycle, the Green Cloud domain must support both current and new business models, even those that have not been invented yet. The concept of microservices and other Green Cloud technologies are "actor agnostic." In other words, the business actors are separate from the technology so that the business model is not hard-coded into its technologies. Business actors can be creative in the use of technologies, supporting innovation and new business models. For example, the VPP is a fairly recent business model innovation made possible by DERs deployed at scale. As more Cloud-enabled solutions enter the marketplace in the coming years, the Green Cloud's technology components will be key to making them a reality.

7.2 Green Cloud Characteristics

A critical enabling concept in the energy IoT reference architecture is the leveraging of reusable componentized services. The rest of this chapter describes the Digital Energy Services Platform (Green Cloud) highlighted with a red border in Figure 7.1. The Green Cloud will:

- Abstract the complexity and brittleness of communications between systems and OT assets using microservices and digital twin agents.
- Provide modern microservices to support virtualization, containerization, and orchestration.
- Leverage messaging bus technologies and payloads such as pub/sub for Cloud-to-edge and Cloud-to-Cloud communications.
- Use rich semantic information model standards to drive interoperability and plug-and-play futures for DERs, gateways, and local controllers.

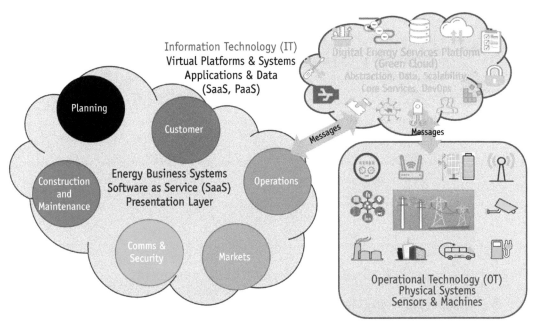

Figure 7.1 Digital energy services platform domain in the energy IoT reference architecture.

- Aggregate/group assets into addressable VPPs to simplify communications and coordination with large numbers of similar assets (by asset type, location, customer type), by point of common coupling.

- Enable intelligent grid-edge devices to self-report by exception or on a timed basis and support an event-driven architecture.

- Simplify data usage with secure data-retrieval services and pub/sub message bus standards to enable easier, less expensive system integration and faster command and control response capabilities.

- Enable a variety of modern data storage systems to support structured, unstructured, smart contract/digital ledger formats.

- Expose energy-specific services for DevOps teams including source code management, software development, and IoT orchestration.

- Eliminate clear text communications and leverage the most current and advanced cybersecurity authentication, authorization, and encryption cipher suite technologies available to address the evolving threat landscape.

- Reduce time and costs for pre-engineering and systems integration by providing services and standardized data definitions.

- Accommodate opportunities to all stakeholders for innovation, new customer services, and energy management capabilities.

- Include forecasting, analytic tools, and other new capabilities to continuously improve and optimize operational and market events.

7.3   Green Cloud Architectural Components

This chapter describes the architectural components within the digital energy services platform in Figure 7.2, which also includes

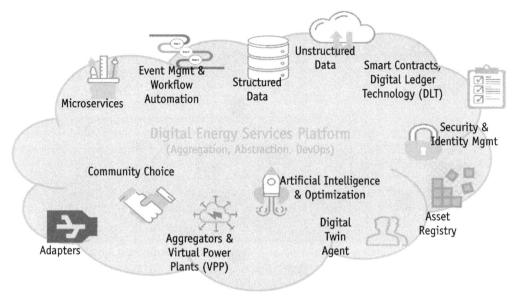

Figure 7.2   Digital energy services platform.

a DevOps environment designed to support the energy software development community. The middle layer in the stack view in Figure 7.3 (briefly introduced in Chapter 4) is the subject of this chapter.

Current electric power business applications often use nonscalable, proprietary pipes to connect business applications. In contrast, the stack view in Figure 7.3 demonstrates business semantics abstraction and hardware-oriented abstraction in an energy services bridge in the middle layer (the Green Cloud domain in the energy IoT reference architecture).

The upper communications layer shows service-oriented architecture (SOA) services in the business semantic layer. The lower layer shows hardware-oriented communications services. Digital twin agents, much like drivers for printers and other peripherals in a computer's operating system, digitally represent physical assets and systems to communicate with adapter services in the OT communications layer.

Siloed systems and siloed data plague the current architectural paradigm. The big "aha" within the energy IoT architecture is to conceptually separate the Green Cloud from the energy systems SaaS domain to directly address the limitations of siloed systems.

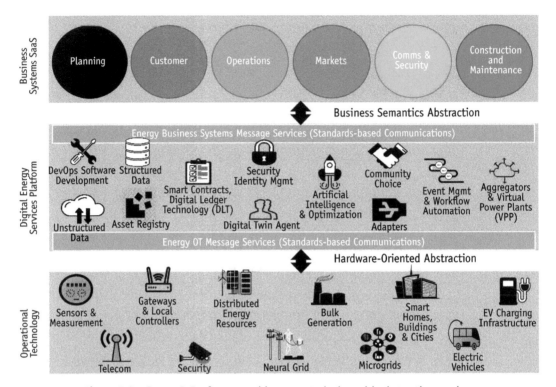

Figure 7.3   Energy IoT reference architecture stack view with abstraction services.

The Green Cloud abstraction layer exists to facilitate seamless communication between physical systems, energy business systems, and data, allowing assets to self-report status data when an event occurs, instead of relying on a polling telemetry system or, really, any system. Notably, modern field devices are intelligent, and their capabilities expand with each new generation. Rather than forcing status reporting through a remote system, this abstraction approach enables grid intelligence by taking advantage of smart edge devices connected to the grid. This approach provides several economic and operational advantages:

1. Data is only reported when threshold value changes trigger an event report, so that only the most valuable information is communicated, stored, and analyzed.

2. Reduced communication traffic.

3. Reduced data storage costs.

4. Reduced computing needs due to reduced data to sort through and analyze.

5. Data is made available directly to authorized subscribers and common data stores through authorized distributed publishers.

Electric power meters offer a cogent example of how such benefits accrue. In today's architecture, in most cases centralized systems poll each smart meter, requesting status information at specific intervals. An IoT, event-driven, data-centric architecture accrues fundamental economic and operational efficiencies.

Imagine a time in the not-too-distant future where a home-owner bids 4 KWh of load shed (demand response) from 5 p.m. to 6 p.m. into a newly created local distribution market. A smart meter recognizes the event or is notified as part of the automated market event workflow. It captures the homeowner's energy usage for the previous hour, automatically reads and records the meter value at the precise moment that the market bid occurs, and does another read at the conclusion of the event. Perfectly correlated historical meter information is available for that event for M&V services to easily be calculated by the market operator and the homeowner or his or her proxy third-party service provider. Settlement for those services could be performed in near real time or, if differences occur between the homeowner and market operator systems, the transaction can be flagged for additional analysis. Further, if the data collected resides in a data store which authorized systems could access, other types of analyses could be performed that provide better event performance forecasts and other operational and economic efficiencies. Such are the fundamental characteristics of a data-centric, event-driven, IoT architecture.

The abstraction achieved by utilizing the Green Cloud unlocks the ability to create a common set of secure energy-specific microservices to access data and use in other systems and by analytic technologies to optimize costs, operations, and other efficiencies. It also enables a distributed intelligence paradigm that leverages smart devices on the grid and in businesses and homes. Finally, a democratized ecosystem emerges to embrace innovation and create an Uber-like environment, providing altogether new economic opportunities for diverse stakeholders working together to deliver more efficient, clean, resilient, and personalized power.

7.4   Cloud Microservices and Container Technologies

Microservices are an API for SOA that decompose software methods into the smallest practical functionality in order to maximize reusability by other services or applications. Microservices and container technologies (modern software techniques offering compelling benefits for the energy industry) hold the potential to virtualize the physical OT assets of the grid, enhancing interoperability between the grid, its systems, and the proliferating DER ecosystem. Microservices and container technologies are the fabric of the Green Cloud's abstraction capabilities.

The Open Group defines a service as supporting four foundational principles [1]. A service:

1. Logically signifies a business activity with a particular outcome for the user.
2. Is self-contained.
3. Is a black box for its consumers.
4. May include additional underlying services.

There are good and bad things associated with microservices. The upside is that using microservices can dramatically reduce the number of development and testing requirements, resulting in faster software development. The drawback to using an individual vendor's microservices is that they can lock in software development teams, who are then forced to rely on proprietary microservices that will not easily port to another vendor's solution.

Containers are the minimum set of operating system (OS) and software application dependency components packaged together required for an application to run as a virtual machine (VM). In other words, containers are an essential tool to perform virtualization and are particularly useful when pushing logic to edge devices, such as gateways and microgrid controllers. Docker is the preeminent tool to create containers that can be deployed to a variety of hardware devices, including servers, personal and industrial computers, as well as single-board computers like Arduino and Raspberry Pi.

Containers really show their magic when it comes to enabling orchestration services. Kubernetes is currently the most popular and widely used orchestration system. It can automate deployments of virtual machines, provide scaling and redundancy, and support automated monitoring and management of container deployments. Some Kubernetes deployments can quickly spin up new or include redundant fail-over VMs if something goes wrong, thereby controlling the overall health and operations of the container. The digital twin agent concept discussed in Section 7.11 is a form of a microservice that may use container technology for virtualization, which would be deployed using an orchestration product such as Kubernetes.

7.5   DevOps Software Development and Source Code Management

The Green Cloud utilizes a DevOps environment to support the software development community, and source code management is a vital element of that. The development environments utilize multiple operating systems, languages, and target deployment scenarios, such as Cloud, on premise servers, hybrid Cloud/premise, and edge devices. DevOps environments include software development services for development, collaboration and reuse, version management, testing, and orchestrated deployments of containers. DevOps environments, supporting microservices, and source code management are included as part of all major Cloud provider solutions.

7.6   Event Management and Low Code Workflow Automation

For this workflow automation discussion, I am not describing the workflow orchestration of containers (e.g., Kubernetes) as outlined in Section 7.4, but rather a simple what you see is what you get (WYSIWYG) or "low code" development toolset. This may or may not be part of the DevOps toolkit described in the previous section. Instead, it is a low code workflow automation tool, such as the visual open-source development environment called Node-RED [2].

Similar to National Instrument's LabVIEW, the Node-RED IoT low code development environment was designed with a 3-layered IoT architecture in mind. Software objects can be dragged and dropped into a software container and wired together to automate workflows in a simple, no-code or low-code way, such as shown in Figure 7.4. Node-RED has hundreds of existing software objects developed and donated by hundreds of software developers within the Node-RED community that enable many different use cases and software deployments simply and quickly.

An event-driven workflow management architecture built on top of node.js, Node-RED can run natively on many low-cost single board hardware platforms, including Raspberry Pi, Beagle-Bone, and Android. As a workflow automation orchestrator, Node-RED simply and seamlessly provides pub/sub message buses (e.g., message queuing telemetry transport (MQTT)). Node-RED applications run within a browser environment and are agnostic to operating systems. At the time of this writing, there were nearly 12,000 open-source contributions, and that number is growing all the time.

Similarly, the Green Cloud would have its own library of vendor-contributed software API objects and workflow automation products to support connecting to their digital twin agents. Additionally, software development companies could contribute by creating new capabilities and exposing innovative services to other developers. While some independent validation and rules will be necessary to achieve this vision, the key takeaway is that tools like the Node-RED low code environment allow for the rapid development and reuse of node objects that will evolve and grow over time. Applications created this way can be implemented using containers and related orchestration services, which was discussed previously in Section 7.4.

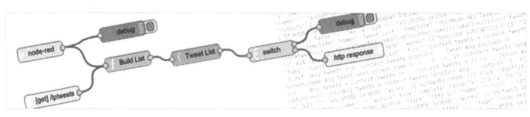

Figure 7.4 Node-RED IoT development tools. (Source: https://node-red.org [2].)

7.7   Data Services

Fundamentally, the energy IoT reference architecture is data-centric. The importance of data cannot be overstated. As stated previously, "it's all about the data!" Being event-driven is also extremely important (the events generate the data) and that is the most fundamental element of the entire IoT ecosystem. Supporting this quintessential principle, the Green Cloud includes a rich set of secure data communication (message bus and payloads) and data storage/retrieval services.

The first and most important step in building the Green Cloud will be designing the correct asset model and time-series data structures and services, which will be difficult and time-consuming. Fortunately, the electric power industry has resources to draw upon, including:

- Rich semantic information models developed by the International Electrotechnical Committee (IEC);
- 61968/70 Common Information Model (CIM);
- 61850 (originally a substation automation standard that modeled traditional assets and that now includes DER);
- IEEE's 2030.5 (Home Area Network, which is harmonized with CIM and IEC 61850);
- OpenFMB (also harmonized with CIM and IEC 61850);
- Other rich semantic information models that are mostly harmonized with these IEC standards mentioned.

Completing this design will require much effort from the experts of these standards organizations. It will also require technology authorities with great familiarity in designing DevOps systems, data repositories, and messaging payloads. The good news is that several commercial solutions already use the IEC semantic models and vocabulary for their internal data schema. Settling on this standard for the electric power industry's schema is a pragmatic mechanism for enabling system-to-system and system-to-data interoperability.

7.7.1   Smart Contracts, Digital Ledger Technology (DLT)

The benefits of distributed digital ledgers are yet untapped in the electric power industry. DLT is often incorrectly called blockchain or cryptocurrency technology. While these are forms of DLT, there are other ways of employing DLT outside of blockchain and other applications. DLT is a legitimate and compelling potential technology for energy applications.

The Green Cloud energy services layer must be founded on data that can be trusted and DLT technologies were designed with trust in mind. Some foundational principles that translate well into an energy IoT data-centric architecture are trustless distributed ledgers and processes, transparent smart contract and business rules, use of encryption algorithms, immutability, and consensus-based auditable and verifiable processes.

7.7.2   Structured Data

Structured data uses Structured Query Language (SQL) or other standardized methods to execute searches and add, modify, or remove fields. This data is extremely organized information stored in fixed fields within relational databases. A common IoT best practice for the present day's relational databases is to create methods or microservices that use encapsulation of standard SQL searches and data-manipulation commands such as RESTful HTTP methods like GET, POST, PUT, and DELETE. This abstraction allows for a highly controlled additional level of access to relational data and simplifies data processing for authorized systems. It also prevents an old, risky behavior of sharing database passwords between developers, jeopardizing table structure integrity and stored information.

7.7.3   Unstructured Data

Information may be generated by humans or machines, and it may be text or nontext data, including videos, documents, contractual information, pictures, logs, and any other type of data that does not fit neatly into a highly structured format. It is important to note that unstructured data is information that does not easily decompose into tables and fields within relational databases.

Cloud providers have mastered the management and storage of unstructured data. Solutions such as Hadoop, Apache Hive, MongoDB, and Cassandra have excellent performance and offer existing tools and services to support highly scalable capabilities.

7.8   Security and Identity Management

The energy IoT data-centricity allow each payload and data process activity to have an extra layer of scrutiny as it travels across systems and data stores throughout the system. Filtering algorithms can check data accuracy to ensure that they are within normal operational limits and can flag data or processes that spoof authorized and authenticated data publisher actors. Analytic tools can be trained to identify potential bad actors, intrusions, theft, and other forms of tampering. Anomalous data or system behavior can be automatically detected, flagged for improper or out-of-limits performance, quarantined from other parts of the system, and could even notify a human to inspect in person. Security is designed in from the outset and leverages the most modern and sophisticated security tools available, providing continuous security improvement, and the ability to learn and adapt using growing data sets as the system matures over time.

As an example, Figure 7.5 depicts one potential opportunity to build an additional layer of security in the energy IoT reference architecture's event-driven, data-centric architecture. The orange-colored objects provide opportunities for security microservices to:

1. Ensure that the data source is authorized and authenticated.
2. Inspect data for unusual values that are not within normal operating limits.
3. Ensure that data is only provided to authorized and authenticated business systems and actors.

Mitigate the occurrence of suspicious data when one or more of the above three conditions are not compliant. Mitigation services could escalate starting by flagging erroneous data and its source, automatically quarantining suspicious data sources, or by notifying and sending a human to perform an in-person visual inspection.

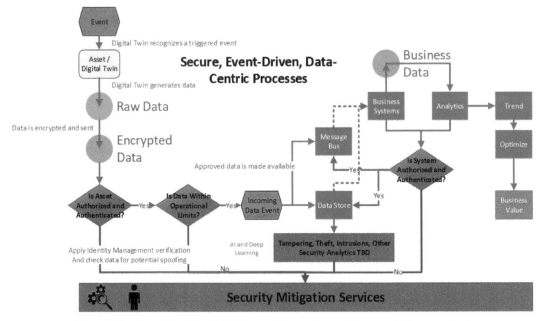

Figure 7.5 Example of how built-in security services could support the energy IoT architecture.

Some electric power industry professionals and many outside of the industry have expressed concern that the large number of DER and intelligent edge devices entering the grid ecosystem will increase the security threat vector surface area and general grid security. Although there is some truth to this, there are other security benefits to the large increases in these intelligent devices on the grid.

First, most DER systems connect to and are hosted in the Cloud, and Cloud providers are recognized as having the most sophisticated security tools available, because their business and reputation demand nothing less. Cloud provider security tools are regularly updated and must address new threats in near real time. Cloud providers have dedicated staff whose sole purpose is to monitor the evolving threat scenario and create processes and tools to mitigate those threats. Deep learning algorithms identify anomalous behavior and microservices to block bad actors as they identify unusual behavior. Cloud DevOps software development environments have security mechanisms in virtually every type of communication. Certificate-based communications are embedded in DevOps-built solutions; similar levels of encryption are used by utilities. Additionally, standards such as IEEE 2030.5 and Open-FMB require encrypted communications in all cases, which is different from some existing SCADA systems that communicate in clear text.

Second, the dramatic expansion in the number of attack vector surfaces is unquestionable. However, the question should be: Is that truly a concern? With so many surfaces in a decentralized system, hackers would need to act at a much larger scale and expend much more effort to create major disruptions than they would hacking a centralized system. Hacking of DER systems would be performed on local systems and cascading events would be preventable. In short, the level of effort required for hackers to disrupt large areas of the grid with decentralization is likely much greater than with a centralized system, making the system more secure the more it is decentralized. Hacking of centralized systems could have larger impacts than those of decentralized systems.

Finally, a trend is that our national security agencies (Federal Bureau of Investigation (FBI), Central Intelligence Agency (CIA), and National Security Agency (NSA)) are opting in to highly secure Cloud-based hosting of their systems and data and out of self-service on-premise IT hosting/management. The implication is that our most cautious and security-focused government agencies prefer Cloud systems, processes, and experts to secure their systems and data over their own IT personnel.

7.9 Asset Registry

Currently, utility business systems typically do not support an asset registry for nonutility-owned assets unless they integrate through the utility's SCADA system. However, in the new world of customer and third party-owned DER, in order to understand what these assets are, their geolocation and capacities, and what capabilities they support, they will need to be stored in the Green Cloud's asset registry. This does not mean that every single asset that a customer owns must be accounted for. Hardware abstractions to BTM assets such as gateways can aggregate those assets within the asset registry. The IEEE 2030.5 standard utilizes gateways in exactly this way, and those vendors may store the individual BTM assets in the gateway itself, or within the Cloud, or both.

As shown in Figure 7.6, Australia has created a national standard for a DER asset registry master file.

The DER asset registry master file is used throughout Australia to track information on energy storage, wind generation, hydro, solar PV, gensets, and EV-charging DER. Australia's Distribution Network Service Providers (DNSP) and the Australian Energy Market Operator (AEMO) use this information to perform forecasts, plan, and to efficiently manage the National Electricity

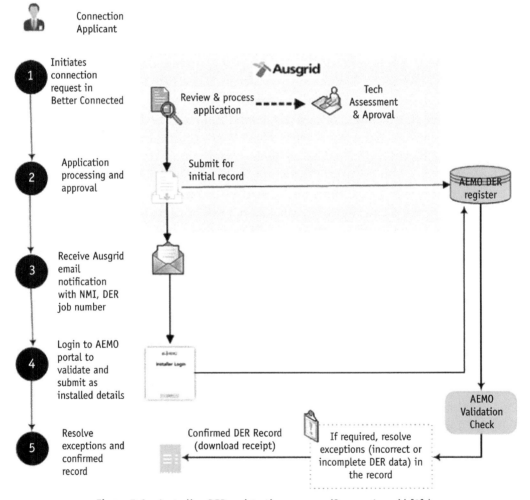

Figure 7.6   Australian DER registration process. (Source: Ausgrid [3].)

Market (NEM) and grid. The asset registry includes the type of DER, its grid location, capacity, and capabilities. Australian rules now require that any new or modified customer electricity generating systems must provide installed DER equipment details. The DER Master File registry can be updated through a vendor's smart device installation tool or by directly interfacing with the AEMO Installer Portal. The United States does not currently have this level of standardization, but could certainly learn from Australia and, in the interest of a common international registry approach, could adopt that standard.

7.10   AI and Optimization

It is an exciting time for the electric power industry and this moonshot opportunity to transform to the energy IoT reference architecture. The ability to leverage this highly flexible architecture to support multiple use cases and employ advanced AI analytics tools for higher efficiencies abound. Section 7.8 discussed the analytics that could be employed to identify bad actors, spoofing, theft, tampering, and intrusions. Similar AI and deep learning methods can be applied for other uses such as grid operations, market participation, grid planning, forecasting, M&V, settlement, and billing services. Within the energy IoT reference architecture, analytics are especially valuable due to the event-driven and data-centric design where the data is meaningful, utilizes reusable microservice APIs, and is only available to authorized systems and actors. As such new analytic tools become available, system vendors will rapidly adopt them to enhance their capabilities, predict and solve problems before they occur, and optimize customer business models. The number and breadth of potential beneficial changes from tools that become faster, better, and cheaper are boundless, making possible an electric power renaissance.

7.11   Digital Twin Agent

The architecture includes an abstraction layer with digital twin agents that simplify the communication between energy systems and grid assets (adapters) and emulate the behavior of assets in simulation environments. This special form of a microservice is critical when scaling grid networks and energy systems to support massive expansion of DER. Digital twin agent adapters can be thought of as containers that spin up when communicating with a physical asset and are removed when communications have completed. Meters and switches are examples of digital twin agent Cloud-based asset communication microservices that come and go only when needed. Some digital twin agents that become real-time, mission-critical pieces of the overall system are likely to remain in memory, physically colocated within or near the asset that they virtualize. This approach can manage digital twin agent container upgrades that provide additional functionality or bug fixes, which can be performed a single time and propagate immediately out to all assets of that model and make.

The architecture envisions agents that announce, describe, provision, and commission the asset. The digital twin agent ensures built-in redundancy/fail-over capabilities. Further, distributed intelligence guarantees that, on failed communication, agents continue to operate independently, tuned to the last command set provided. However, the lack of an energy standard for digital twins makes this vision a technology gap that needs to be addressed.

7.11.1   There Are Probably at Least Two Types of Digital Twins

Figure 7.7 describes my current conception of digital twin agents, which is still evolving. More work will be needed to establish agreement and common interfaces, features, and containerization/orchestration methodologies. That said, technologies such as Docker and Kubernetes already have many of the capabilities to support the functionality envisioned for digital twin agents in the real world.

Many people believe that a digital twin is a virtual mirror or emulator of the physical hardware or entity being represented, one that accurately mimics the asset's operations and behavior. I chose the term "agent" as part of the name to help the reader recognize that there is still much discussion and unclear definitions for the digital twin. Whatever the definition becomes, Figure 7.7 provides some baseline requirements and optional requirements for the digital twin agent concept.

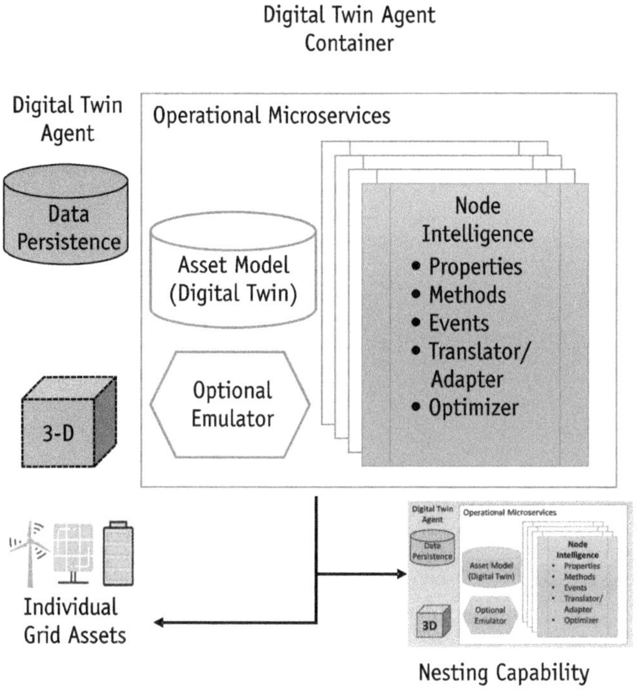

Figure 7.7 My digital twin agent concept.

Table 7.1 provides my concept of minimal functionality requirements for the digital twin agent. In some IoT implementations where digital twin emulators support simulation, a digital twin agent may be used in power flow modeling or planning exercises. In this case, the digital twin is quite sophisticated and requires an operating environment with significant horsepower (i.e., a server) to mirror the full behavior of the physical asset that includes an emulation engine.

Another type of digital twin agent, a communication abstraction, provides services and systems with a bridge to an asset that standardizes communication to assets of the same class through adapters translating semantic message payloads regardless of the protocol the hardware speaks (e.g., Modbus, OPC, DNP). Some IoT developers may call these adapters, rather than a form of digital twin. Regardless of the naming convention, a digital twin agent acting as an adapter is far more, including such functionality as an event handler, archivist, supported properties and methods, data persistence, data filtering and cleansing, and optimizer. However,

Table 7.1
Digital Twin Agent Concept Minimal Functionality Requirements

Requirement	Minimal Functionality	Comment
Can be located anywhere	Yes	Can be located in the Cloud, near the asset, or as part of the asset
Persists asset information	Yes	The digital twin must support data persistence to provide at least short-term historical data for other systems and digital twin agents
Event-driven architecture	Yes	The digital twin agent must support methods to react to certain events that may include executing an algorithm, triggering another event for other system actors, or sending/storing data for historical purposes
Emulation	No	When performing simulations of assets or systems of several assets, emulation is necessary; however, my concept of the digital twin agent may act as an intermediary to perform optimization or event management functions where emulation is not required
3-D model	No	As an option, a digital twin agent may include a 3-D model that could be used to support installation and maintenance activities
Use of orchestration services	Yes	The digital twin agent is a containerized application and can be orchestrated using services such as Kubernetes
Has intelligence	Yes	Includes properties, methods, events, optimization algorithms, and asset communication capabilities
Point of aggregation	Yes	Can behave as a point abstraction and support nesting functions for assets and other agents; this provides fault tolerance and the ability to operate autonomously with other assets
Low code technology	No	Low code technologies are optional

this second, less sophisticated type of digital twin does not do emulation.

Digital twin agent adapters have fairly low processing power needs, they can be physically located in the cloud or on the OT physical system near the asset and are compact. For the purposes of the energy IoT reference architecture, adapter functionality is a minimum requirement of a digital twin agent. Having additional emulator capabilities would be helpful when performing simulation and optimization processes.

7.12 Aggregators and VPPs

Customers and third-party assets are already becoming part of the smart grid landscape. An aggregator may serve the power consumer by uniting customers and negotiating on their behalf to coordinate with a utility's customer programs, as well as managing the customer's local energy assets. For example, PV solar installation companies may connect aggregated customer systems through their Cloud to offer services to the utility. The aggregator may also combine multiple buildings into VPPs and negotiate on the customer's behalf to sell grid services, participate in markets, or provide other services to the utility or third parties.

VPPs, aggregations of DER that provide distributed generation, controllable loads for DR programs, and/or energy storage capabilities, are a compelling solution for both commercial vendors and utilities alike. Quickly gaining popularity, VPPs are providing DER aggregators new business opportunities in markets and in providing for-fee grid services, as shown in Figure 7.8..

7.13   Community Choice Aggregation

In some states, community choice aggregation allows local municipalities to opt out of their regulated utility and negotiate in bulk markets to become the service provider for local community elec-

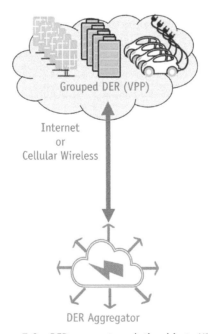

Figure 7.8   DER aggregator relationship to VPPs.

tricity needs. This is a different kind of aggregator than described in Section 7.9. CCAs are driven to support customers with low carbon, clean energy solutions and negotiate with third-party wholesale energy providers to purchase electricity within a defined jurisdiction. Regardless of what drives it, be it cost savings or local economic incentives or policies requiring greener generation portfolios, this trend will likely continue. The architecture endorses the need for nonutility entities to participate within the OT domain as either electricity producers or consumers.

7.14   Adapters

The simple idea of using adapters as translators to convert one protocol language to another gets more complex in practice, especially when communicating with proprietary systems or registry-based systems such as Modbus, CAN bus, and OPC-UA. With myriads of asset types and vendors and their unique interpretations of protocol intent, mistakes will happen. However, adapters have become critical to the electric power industry ecosystem. Many gateways now include a universal translator capability to simplify communications between business systems and the variety of communication protocols used by energy equipment and local energy management systems.

7.15   SOA, Message Buses, and Message Payloads

The general idea behind SOA is the concept of loosely coupled services that deliver some business value and that are reusable by other services and applications. This simple concept that arose from the early 1990s with the advent of Visual Basic eXecutables (VBX) and ActiveX reusable components allowed software developers to embed highly complex capabilities into their own applications. Components' events, properties, and methods abstract the inner logic as a black box. Conceptually, SOAs provide the same loosely coupled foundation.

Message-oriented middleware or message buses broker communications between different assets, services, and applications. Message buses commonly include message queues and intelligent brokers that route communications quickly and efficiently. Many message buses can even guarantee message delivery through quality-of-service functions.

Some message buses route messages in broadcast or multicast modes to allow any approved actor on the bus to see the message. While not optimal for security, this method also creates additional processing needs, as each actor on the bus needs to determine whether the message was intended for them or not. Instead, the pub/sub message bus technique adds an extra layer of efficiency by using topics, where authorized publishers publish topics only to authorized subscribers. This approach yields a much more structured and granular management of data communications, allowing only actors that can send that data (publishers) and is only routed to the actors that can receive it (subscribers). Using pub/sub messaging services, peer-to-peer data routing is performed to target the correct actors providing and requiring specific data that can be encrypted for additional security.

Message payloads can be dispatch schedules, time-series data, or any form of structured and unstructured information. Payload sizes can be very small, especially when using modern compression and binary technologies, allowing low latency communications. In order to facilitate better interoperability between systems and devices, agreeing on common messaging and information model standards is critical. Standards organizations sometimes develop information models that support specific sets of use cases and start from larger international information models such as CIM and IEC 61850. These profile standards promote better interoperability by using the same asset data models and vocabulary. OpenFMB and IEEE 2030.5 are two good examples of profile standards.

Loose coupling and message-based communications underpin the energy IoT reference architecture's principal goal for interoperability. SOA, message bus technologies, and common message payloads promote better interoperability and more rapid integration of new systems and grid assets.

7.16 Conclusion

The digital energy services platform domain (Green Cloud), the heart of the overall energy IoT ecosystem, is the highly scalable abstraction layer that dramatically simplifies integration of new assets and systems. The microservices provide black boxes to access data, employ analytics and deep learning techniques, interoperate with other applications/services, utilize digital twin agents to connect to grid assets, and implement virtualization, containerization, and orchestration.

The Green Cloud leverages existing Cloud provider environments but extends those existing generalized services with energy-specific IoT microservices to support the specific needs of electric power industry users. The Green Cloud layer of the energy IoT reference architecture is flexible, highly scalable, adaptable, extensible, and secure. It uses the Cloud provider's native environment to provide the most advanced security and identity management technologies available. The Green Cloud extends the Cloud provider's DevOps environment, providing common energy-specific asset data models and vocabulary, abstraction through the use of pub/sub messaging services and adapters, and other microservices that reduce engineering, software development, and system integration time and costs. The architecture is data-centric and event-driven, communicating and storing only the most important data when things change or at some time interval. This reduces communication traffic and the amount of data to inspect, store, and analyze. Development of the Green Cloud layer of the energy IoT architecture will not only financially benefit the companies that develop it, but will also provide society the mechanism to modernize the electric power industry's technology. The Green Cloud will provide a method to directly address the reduction of GHG emissions and enable the rapid integration of more economical generation sources such as utility-scale PV. Creating the Green Cloud Layer is nothing less than a moonshot opportunity for mankind.

References

[1]   The Open Group, "Service-Oriented Architecture – What Is SOA?" https://www.opengroup.org/soa/source-book/soa/p1.htm.

[2]   Node-RED, https://nodered.org/.

[3]   About the Distributed Energy Resources Register, https://www.ausgrid.com.au/Connections/solar-battery-and-embedded-generation/Distributed-Energy-Resources-Register, Ausgrid.

8

Mapping the IEEE 2030.5 Protocol to the Energy IoT Reference Architecture

This chapter is the first of two chapters focused on bringing the energy IoT reference architecture theory discussed in the last four chapters into real-life practice.

Mapping a standard or organization's architectural best practices to the energy IoT reference architecture (or any reference architecture) can be difficult, especially if the number of architectural layers in the standard or best practice does not allow for a clean mapping to the three-layer architecture described in Chapters 3 to 7. Therefore, when building the solution architectures that will be illustrated in Chapter 9, this standards-mapping exercise is optional for readers not familiar with technical standards.

8.1   History

The IEEE 2030.5 standard was originally part of the Zigbee standard and was called the Smart Energy Profile (SEP). That standard was intended to use the meter as a gateway into homes and businesses with the Zigbee wireless protocol as communications transport. In 2009, the National Institute of Standards and Technology (NIST) and the Smart Grid Interoperability Panel (SGIP) identified SEP as a critical Smart Grid standard. However, this standard still had issues, as it intermingled architectural layers, so it was placed on a fast track to separate the information model from the transport layer. That effort yielded SEP 2.0 and was developed with solid engineering and thoughtful industry collaboration, but SEP 2.0 never really took off until the IEEE decided to run the standard through their internal processes and ratified it in 2016.

IEEE 2030.5 is based on the CIM that was developed by the IEC standards organization. In the 2018 update to the standard, the IEEE 2030.5–Smart Energy Profile Working Group adopted the DER information model used in the IEEE 1547-2018 interconnection standard for DER, the same DER models defined in IEC 61850. Using these common asset models makes interoperability much more achievable, which is a problem with which the electric power industry has struggled. This rich semantic information model standard allows for message-based communications rather than old telemetry register-based communication, so no register mapping is required.

The concept of grouped or aggregated assets and hierarchical points of coordination was designed into the IEEE 2030.5 standard. This allows the standard a great deal of flexibility and the ability to support a wide range of IoT use cases including DER visibility, situational awareness, and coordination. FERC Order 2222 now requires that all TSOs provide DER aggregators with market participation opportunities. Aggregation capabilities are inherently baked into the IEEE 2030.5 architecture using physical gateways to aggregate DERs and that can even nest with other gateways in parent-child hierarchical relationships.

IEEE 2030.5 is one of the few true energy IoT standards developed for modern grid communications, distributed intelligence, and interoperability.

OpenFMB is another IoT standard with incredible implications to support a nimble and neural grid as it matures. A short description of the OpenFMB standard is included in the appendix.

8.2   IEEE 2030.5 Architecture

The architectural components of the IEEE 2030.5 protocol shown in Figure 8.1 are:

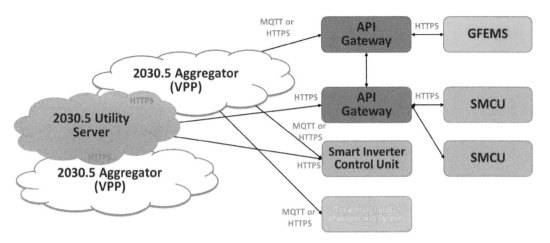

Figure 8.1   IEEE 2030.5 architecture.

- *Utility server:* The 2030.5 utility server, located in the Energy IoT Energy Business Systems SaaS layer, communicates with other utility business systems to provide information such as visibility, situational awareness, and available capacities. The utility can perform queries, negotiation, and dispatch of aggregated DER by communicating via a RESTful protocol with the Digital Energy Platform Services Domain (the Green Cloud)-based IEEE 2030.5 Cloud Aggregator. Notably, the name "utility server" is not exactly accurate, as the utility server could also be used by a distribution system operator or a VPP company managing many DERs (or EV chargers or microgrids) to coordinate and optimize for economic or other business reasons.

- *Cloud aggregator:* The Cloud aggregator is hosted in the Green Cloud. Aggregation is another method of abstraction, grouping assets into a single interface with which other systems can coordinate. This is another black box abstraction that does not require direct communication with individual assets or even knowing what those assets are. You can consider it a form of a VPP that coordinates capacity and capabilities across numerous DERs. Both the Cloud aggregator and the 2030.5 edge gateway can support the MQTT pub/sub protocol, so communications between them are event-driven. There are no client/server communication requirements that require the client to check in with the server for new messages, which can further streamline communications requirements.

The Cloud aggregator, the intelligent piece of the 2030.5 architecture, monitors all the gateways and/or Smart Inverter Control Unit (SMCU) and generation facility management system (GFEMS) assets. Further, it groups those assets into logical units, manages the DER system of record, updates available capacities, and provides dispatch schedules to the gateways, SMCUs, and GFEMS based on the utility server's requests. The Cloud aggregator monitors dispatch events and collects behavior and performance information to support analytics and other metrics. Finally, the Cloud aggregator includes a DER registry system of record that can be leveraged by the utility to support grid planning and even grid connectivity and power flow models.

- *2030.5 gateway:* The 2030.5 gateway is an abstraction to the OT physical layer. It is the working-class device of the 2030.5 architecture. The gateway provides several different capabilities to simplify communications and aggregation. Not only can it act as an aggregation point for any DER with which it can communicate, it can also discover IEEE 2030.5 devices. Gateways can provide flexibility for layered hierarchical architecture systems designs by aggregating other gateways. They may also act as a universal translator that transcribes IEEE 2030.5 messages into native DER protocols such as Modbus, DNP3, CAN bus, and IEC 61850, among others, and it can be programmed to speak to NEST, ECOBEE, and other proprietary vendor protocols. Lastly, additional capabilities can be installed remotely by the IEEE 2030.5 aggregator using containers and orchestration while also offering an EMS option that allows the customer to manage the DER assets on their own premises.

- *GFEMS:* The GFEMS is the local controller for a facility. The facility could be a building or another type of intelligent controller that manages/controls several energy devices such as a building management system. The 2030.5 gateway and GFEMS could be housed in a single system.

- *SMCU:* The SMCU supports AC to DC electricity conversion—a smart inverter. Most importantly, it is also the communications interface and brains for controlling PV solar, batteries, and EVs.

8.2.1   CPUC Rule 21

In 2019, the CPUC created the Rule 21 mandate for integrating behind-the-meter DER. The ruling was primarily directed at rooftop solar that was being installed at record pace. The order is now a requirement for all new construction in California. The ruling requires that all inverters connected to the grid must comply with IEEE 1547 interconnect requirements and one of three communications protocols (Figure 8.2): IEEE 2030.5, SunSpec Modbus, and IEEE 1815 (often referred to its former name, DNP3).

The CPUC created Rule 21 to ensure safe and interoperable use of DER. As shown in Figure 8.3, five key drivers inspired their decision to construct the legal requirement that all grid-connected DER that perform export meet these communications standards requirements.

California is committed to decarbonization and recognizes key problems faced by new stakeholders, IoT assets, and intermittent generation and unanticipated loads that must be addressed to be successful in its decarbonization ambition. DERs must increase resilience while maintaining current safety and reliability requirements. Nonutility-owned DERs will need to support utilities and grid stability by providing grid services individually or through aggregation. New stakeholders are welcome and will be encouraged to participate in grid services through utility customer programs. New stakeholders may also participate in wholesale and local markets. There is a need to simplify costs for integrating DERs

Figure 8.2   CPUC's three allowed communication interface standards.
(Source: SunSpec Alliance [1].)

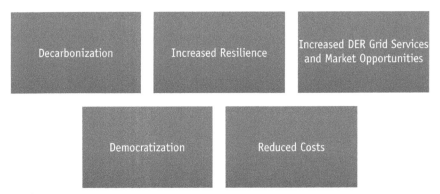

Figure 8.3   CPUC drivers for DER interoperability in California.

into utility operations and markets so that they are cost-effective to execute.

As the default standard for DER integration in California, the smart inverter vendor community is quickly getting up-to-speed with implementing this interface capability so they can sell their products there. However, as this standard becomes the norm for smart inverter manufacturers, it will likely drive other states and countries to mandate IEEE 2030.5 as one of the accepted protocols to use within their territories. It will also create an ecosystem for DERMS and VPP vendors to automatically communicate with verified, authenticated, and authorized inverters, driving these types of consumer devices closer to a plug-and-play capability. IEEE 2030.5 is gaining traction and will be an important standard in the days and years to come.

8.2.2 IEEE 2030.5 Features and Supported Grid Services

Developed with a rich set of features and capabilities, IEEE 2030.5 was designed to support use cases related to nonutility-owned DER assets to coordinate with utility operations, primarily focused on behind-the-meter assets. It continues to evolve to support even more use cases, efficient bidirectional communications between the Cloud and edge devices, and continuous modernization of security best practices to match evolving threats. Formalized testing and certification remain available to ensure compliance with the standard.

8.2.2.1   Semantic Information Model

IEEE 2030.5 was developed to support DER integration and the ability of DER to support a variety of grid services. It is a client/server architecture designed to support all forms of DER: flexible controllable loads, distributed generation, energy storage, and EV charging. The standard leverages a rich semantic information model based on the IEC's Common Information Model (IEC 61968) to provide numerous grid service messages. Messages are XML or Efficient XML Interchange (EXI) constructs using a schema that is derived from the CIM protocol UML model. This makes the IEEE 2030.5 information model a profile or subset of the CIM model with some extensions.

8.2.2.2   Publication/Subscribe MQTT and RESTful Communication Transports

The RESTful (HTTPS over TCP/IP) and subscription/notification (such as MQTT), communication protocols used by IEEE 2030.5, are well-understood and in common use by software developers worldwide. No licensing or usage fees are needed for MQTT or HTTPS. Both transport mechanisms are natively supported by platform DevOps and other commercial software development environments. The RESTful interaction model supports GET, PUT, POST, and DELETE commands. Both IPv4 and IPv6 Internet Protocol (IP) versions are supported by the standard. The standard is agnostic to physical transport communication and can use any underlying communication transport that uses the IP including Wi-Fi, ZigBee IP, WiSUN, CellNet, and HomePlug.

Communications with Smart Inverter and GFEMS end devices are performed through secure RESTful HTTP. HTTP is a low-risk implementation as it is used extensively by developers as internet web services to communicate between applications. Any device can be a server or a client, so edge assets can self-report and communicate with other DER assets and the Cloud aggregator or utility server when reportable events or issues occur. Gateway and Cloud aggregation devices that communicate both upstream and downstream may implement both client and server capabilities, which requires significantly more coding than when implementing client functionality only.

One challenge when using HTTPS over the public internet is that latency can be an issue, but this can easily be overcome because dispatchable DER assets are typically scheduled (normally at intervals under 15 minutes) for the next 24 hours. Subsecond communications to support highly dynamic grid services are typically not needed from most DER assets other than the battery, and even batteries can be sent messages to function in an operational control mode that automatically provide dynamic services such as frequency, voltage, and VAR regulation support. However, one of the guiding principles defined in California's Rule 21 is that DER systems are not intended to support subsecond interactions and protection.

IEEE 2030.5 utilizes HTTPS between the server and Cloud aggregator levels, with HTTPS or MQTT being optional between the Cloud aggregator and the gateway. The use of the MQTT pub/sub communication protocol supports rapid, guaranteed message deliveries and configurable QoS parameters. MQTT is a lightweight messaging transport that provides a simple mechanism for creating topics that allow authenticated and authorized devices to publish or subscribe to data within the topics. The overall intent of utilizing MQTT topics is to provide efficient, event-driven, bidirectional communication between the Cloud and edge devices, creating a simple way for groups of devices to interoperate with one another.

8.2.2.3   Cybersecurity

Chapter 7 discussed cybersecurity and the challenges of introducing large numbers of DER that result in a dramatically increased overall threat surface for bad actors to exploit. IEEE 2030.5 was designed with security built in from the beginning.

IEEE 2030.5 provides a highly secure certificate-based communication capability, as shown in Figure 8.4, which currently requires HTTPS/TLS 1.2 and mandates a single cipher suite using a 128-bit security level. The current recommendation is compliant with NIST SP 800-57 and NSA Suite B Recommendations for Cryptographic Key Management. As new versions of TLS cipher suites evolve, the 2030.5 standard will no doubt conform to those newer security requirements and best practices.

All certified products require a certificate issued by the IEEE 2030.5 Certificate Authority (CA). The certified public key infrastructure (PKI) program is managed by the SunSpec Alliance to authenticate and secure communications using Elliptic Curve

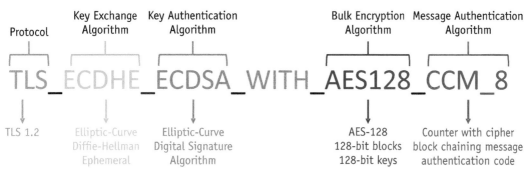

Figure 8.4 IEEE 2030.5 cipher suite compliant with NIST and NSA security recommendations.

Cryptography (ECC) between clients and servers in the IEEE 2030.5 ecosystem. Current device certificates never expire and are irrevocable, but there are discussions to change this in future versions of the standard. The use of this PKI program is a requirement for product certification, and the program is designed to increase confidence in the overall security of communications between DER devices and the communications between the DER devices and the Cloud aggregator. As part of the IEEE 2030.5 certification process, DER assets must pass standardized secure communication tests that utilize test PKI certificates. Once a DER device has passed certification, the vendor is issued production PKI certificates for that device product.

8.2.3   Discovery

One feature that truly sets IEEE 2030.5 apart from other protocols is its ability to discover other IEEE 2030.5 architectural components and DER assets on the network. The standard uses the multicast Domain Name Server (mDNS) protocol for host discovery and the DNS-SD (Service Discovery) protocol for resource discovery on networks that are enabled with the zero configuration standard. The standard can also use the standard DNS protocol for host discovery. In essence, IEEE 2030.5 creates a self-configuring network of DER energy assets from the home area network (HAN) to the utility server with existing networking protocols, increasing the possibility of true plug-and-play as the electric power industry transformation matures.

8.2.4 Supported Grid Services

Electricity supply and demand must be balanced. Electricity reliability and power quality must also be maintained on the grid with proper voltage and frequency. Historically, this balance of energy and capacity and proper power quality (called essential reliability services (ERS) or ancillary services) has been managed through bulk energy resources using market systems that benefit the grid and the grid service suppliers. DERs provide an opportunity to utilize distributed assets for these critical grid services, and IEEE 2030.5 was designed to support energy, capacity, and ancillary services.

Figure 8.5 shows NREL's highly publicized graphic of primary grid services broken out by energy and capacity and ancillary services, with associated timescales. Numerous related DoE publications are available at no cost online. In general, grid services are services that:

- *Provide energy shifting and capacity services:* These services shown at the top of Figure 8.5 ensure that adequate generation capacity is available to support demand needs. DERs

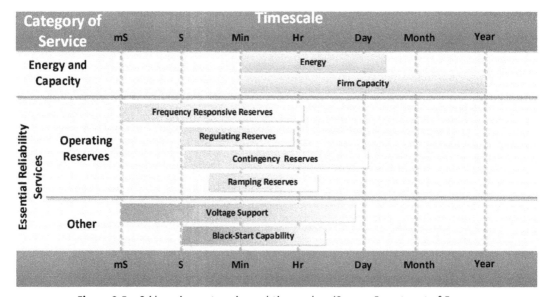

Figure 8.5 Grid services categories and time scales. (Source: Department of Energy National Renewable Energy Laboratory [2].)

can help to provide these services in peak demand conditions rather than firing up expensive peaking plants. These services can be achieved through flexible loads and DR programs or by leveraging distributed generation, or both.

- *Provide reserve services:* These services provide a backup plan in case of an unexpected outage or other grid asset failure. DER assets can be used to support the grid with additional generation, load shedding, frequency and voltage regulation, and ramping capabilities when other parts of the grid fail or cannot match grid needs without additional reserve resources.

- *Allow for grid upgrade deferrals:* Like energy shifting and capacity services, grid assets are used to provide energy and capacity rather than incurring expensive grid investment costs to meet peak demand needs. Because peak demand conditions only occur occasionally at predictable times of the year, DERs can be used to match the additional capacity and energy needs for those times.

- *Provide fast response ancillary services:* These services include synchronized regulation with voltage, frequency, and VAR support. DER can be used to support highly localized fast response ancillary services. A good example of this type of service are flywheels that can respond very rapidly.

- *Provide black-start capabilities:* When the grid fails and the system needs to be restored, assets that can provide grid-forming capabilities that can create the electrical sine wave to allow other assets to synchronize to are needed. Most DER assets do not have this capability and are grid-following assets. However, microgrids and some smart inverters for batteries are beginning to support this grid service need.

- *Maximize clean energy opportunities:* Although this is not a typical grid service yet, an opportunity presents itself to leverage clean energy assets such as renewables and batteries to reduce carbon emissions.

The use of DERs to provide grid services is a growing area, and, as capabilities expand, DER will displace more traditional bulk energy assets. Service providers must learn proper coordination and how to truly leverage the special DER capabilities as they become the dominant asset for grid services. IEEE 2030.5 provides functions to support most of these services and will continue to add new functionality with each new upgrade.

IEEE 2030.5 adopted the IEC 61850-70-420 standard concept of logical node classes for DERs, which group different functions into function sets or toolsets that include:

- Time synchronization;
- Smart inverter settings messages;
- Pricing messages;
- Security;
- DR and load control messages;
- Device capabilities' information;
- Energy usage metering information;
- DER program information;
- Billing information;
- Prepayment metering messages;
- EV-charging reservations.

Additional functions support file downloads/updates/bug fixes and service provider messages such as weather alerts, tips, general messages, and Amber Alerts.

As stated previously, IEEE 2030.5 was primarily designed to support grid services for utility operations. The function sets, shown in Figure 8.6, provided in IEEE 2030.5 gives modern, message-based DER communication methods that can scale with the rising levels of DER penetration. As more DER-friendly systems mature, IEEE 2030.5 will gain traction with utilities, aggregators, and vendors to provide more meaningful grid services from DER regardless of ownership.

8.3   IEEE 2030.5 Certification and Testing Tools

The IEEE 2030.5 ecosystem includes a rich set of certification resources and testing tools. Conformance test tools are available

Figure 8.6 IEEE 2030.5 functions.

from QualityLogic, and the IEEE 2030.5 Common Smart Inverter Profile (CSIP) certification is led by the SunSpec Alliance.

8.3.1 Test Tools

QualityLogic's test tools provide functional testing software for IEEE 2030.5 clients and servers. They provide test kits for vendors to conform with the entire IEEE 2030.5 technology stack. It includes a DERMS simulation that facilitates the California Rule 21 Testing Pathway (CALSSA) for integration with the IEEE 2030.5 utility server. This includes interoperability testing of autonomous functions, communications protocols, and advanced functions defined in Rule 21. These functions include metering, pricing, DR, and Smart Energy testing capabilities.

Other available test kits address core functionality of IEEE 2030.5, including basic functions, communication, aggregation, and error handling. They address all the tests defined in the

SunSpec California Rule 21 Test Procedures for the Utility Server, Cloud Aggregator, and End-Client (gateway and inverter) tests.

QualityLogic provides a free application guide that provides instructions for performing CALSSA testing using their tools.

QualityLogic's suite of test tools allows vendors to create IEEE 2030.5-conformant products that will assist them in and accelerate the overall certification process. Vendors can test a variety of use cases and scenarios that can be implemented using real DER assets or simulated devices. APIs are included that can be integrated with the vendor's regression test system tools for simpler, more automated testing processes. Testing tools can be purchased directly from QualityLogic.

8.3.2 CSIP Certification

The IEEE 2030.5 standard includes an accompanying certification process and set of testing standards led by the SunSpec Alliance. Certification is for specific application profiles of IEEE 2030.5. The certification process includes testing for compliance with California Rule 21 specifications. SunSpec has partnered with numerous global authorized test labs (ATL) to provide independent third-party certification testing. ATL test engineers perform CSIP testing using the same QualityLogic test suite. The ATLs provide test results to SunSpec for evaluation for compliance with SunSpec CSIP and proper execution of SunSpec's CSIP test procedures.

Certification processes support the entire IEEE 2030.5 technology stack: utility server, Cloud aggregator, smart inverter, and gateway products. Upon successful completion of the independent certification process, vendors are awarded a security certificate and can include the logo with their product(s) (Figure 8.7). Strong authentication and encryption requirements are a key aspect of the certification process. Certified vendor products are also provided with SunSpec's PKI certificate to enroll in the IEEE 2030.5

Figure 8.7   SunSpec Alliance's certification logo for the Common Smart Inverter Profile. (Source: SunSpec Alliance [1].)

authentication system. SunSpec recently announced the development of a similar security suite and certification pathway regionalized for Australian rules.

The SunSpec Alliance provides numerous supporting documentation, specifications, and resources to support vendor CSIP
certification. The details of SunSpec's certification process can be
located on their website.

8.3.3 IEEE 2030.5 Architecture Mapped to Energy IoT Reference Architecture

As mentioned previously, a common IoT pattern is shared by the
IEEE 2030.5 architecture and the energy IoT reference architecture.
Figure 8.8 shows the alignment with the business systems SaaS,
digital energy services platform (Green Cloud), and grid asset OT
layers. The IEEE 2030.5 technology stack includes a 2030.5 utility
server with SaaS systems running at the utility or system operator. IEEE 2030.5 also has a Cloud-based aggregator and a physical
piece of edge hardware called a gateway that acts as an abstraction
layer, aggregating the local DER assets and creating a VPP for simplified and more meaningful interoperability options. The gateway can support a variety of asset communications and functions
and can also be virtualized in the Green Cloud for some use cases
and implementations. This type of architecture can be deployed in
numerous configurations and different levels of hierarchy and abstraction. In theory, there can be multiple aggregators and multiple
gateways nested as deeply as needed.

The IEEE 2030.5 three-layer architecture is highly flexible. It
was designed to utilize standard pub/sub and RESTful communication protocols with a rich semantic information model that
is a subset or profile of the Common Information Model (IEC
61968/70). As you can see from Figure 8.8, because IEEE 2030.5
was designed with IoT in mind, the mapping of it to the energy IoT
reference architecture reveals a simple 1:1 relationship:

- IEEE 2030.5 utility server *maps to* the energy IoT business
  systems SaaS layer.
- IEEE 2030.5 Cloud aggregator *maps to* energy IoT digital energy services platform layer (Green Cloud).

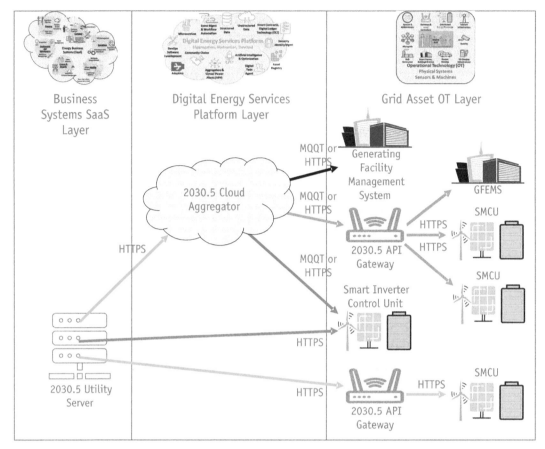

Figure 8.8   Flexible communications with DER assets using the IEEE 2030.5 technology stack.

- IEEE 2030.5 API gateway, SMCU, and GFMS *map to* the energy IoT grid asset OT layer.

8.3.4   Where to Find IEEE 2030.5 Documentation

The IEEE 2030.5 standard is available for purchase for $280 (discounted for IEEE members) at the IEEE.org website [3]. Test tools are available from QualityLogic at their website [4]. Certification information, test PKI certificates, and authorized test lab details can be located at the SunSpec website.

8.4   Conclusion

IEEE 2030.5 was designed as an energy IoT standard from the outset. Its highly flexible design and rich set of functions simplify communications with individual and aggregated DERs of all types and it maps cleanly to the energy IoT reference architecture.

Although the standard was originally established in 2007 and has been through several version updates, the increased adoption of IEEE 2030.5 is a more recent event and is driven by IEEE sponsorship and the CPUC's California Rule 21 mandate. Work is in progress to add new capabilities for more comprehensive and sophisticated smart inverter use case functions. Work is also underway to integrate more simply with EV infrastructure. As vendor adoption and native support for the standard increase, we can expect to see other states and territories establish mandates like those in California.

The standard uses low-risk and well-understood pub/sub MQTT and RESTful HTTP communication transport standards. It uses the NIST TLS security recommendations that employ PKI certificates for all certified DER products. The architecture is client/server and highly versatile: any device can be a server or a client, so DER assets can self-report and communicate with other authenticated and authorized DER assets and the Cloud aggregator or utility server. Conformance and interoperability testing tools can be purchased from QualityLogic. The SunSpec Alliance manages the CSIP certification process and has several worldwide testing facility partners to support the certification process.

The standard has numerous free artifacts for download that can be found at IEEE in an internet search, including the schema and sample code. The standard can also be purchased in electronic and printed formats at the IEEE Standards Store with discounted prices for IEEE members.

References

[1]   SunSpec Alliance, https://sunspec.org/certification/.

[2]   Denholm, P., Y. Sun, and T. Mai, "An Introduction to Grid Services: Concepts, Technical Requirements, and Provision from Wind," National Renewable Energy Laboratory (NREL), January 2019, https://www.nrel.gov/docs/fy19osti/72578.pdf.

[3] IEEE Standard for Purchase, https://standards.ieee.org/standard/2030_5-2018.html.

[4] QualityLogic Suite of Test Tools, https://www.qualitylogic.com/what-we-test/smart-energy-standards/ieee-2030-5-test-tools-qa-services/.

[5] SunSpec Alliance 2030.5 certification information, https://sunspec.org/common-smart-inverter-profile-csip.

9

Developing Energy IoT Rapid Solution Architectures

Chapters 1 to 3 provided the reader with the challenges of climate change, traditional technology, and architectural challenges and why transformational change is needed, and, like it or not, change is coming, faster than we think. Chapters 4 to 7 offered the reader the theory behind the energy IoT reference architecture, providing a high-level overview of the three-layer IoT architecture and deep dives into the architectural components of each domain.

A thorough understanding of the energy IoT reference architecture must go beyond these foundational chapters to reveal the many good ideas that, unfortunately for now, remain theoretical. For instance, the focus of this chapter, developing energy IoT rapid solution architectures, provides the reader with a methodology and approach that will turn theory to real-life practice.

Solution architectures are highly focused architectures for IT products that support one or more specific business use cases. In other words, solution architectures are designed to help organizations to solve one or more business challenges. Many solution architectures start on a whiteboard. Simple rectangles, lines, arrows, stick figures, and circles depict functional components, relationships, actors, and data. These often-collaborative brainstorming sessions among a few individuals (2 to 3) armed with colored markers begin the process of crafting visual representations to show how one or more solutions might support use case(s) and solve

challenge(s). These types of sessions often die right where they begin, on the whiteboard, probably more often than not. Sometimes
the solution idea fails to progress into an actual product development effort for something as trivial as poor drawing skills or from
an overly complex representation of the solution. Perfection is not
required at this stage. Simple is better.

The emerging energy IoT landscape already provides so many
opportunities for innovation and use case-driven solutions. Many
challenging problems already require new thinking and creative
ideas from the most unlikely places or people. Communicating
these ideas and innovation concepts is difficult for a start. Drawings are the easiest way to express a new idea in a language that
lets stakeholders easily understand, critique, and enhance. This
chapter discusses a simple methodology and provides an example for creating a simple solution architecture that is easy for both
technical and nontechnical people to build and to explain to other
stakeholders.

9.1   Developing Energy IoT Rapid Solution Architectures

This chapter focuses on providing the reader with a simple solution architecture development methodology.

9.1.1   Energy IoT Rapid Solution Architecture Methodology

Just about anyone could be a solution architect, which may be a bit
of a shocking statement. By no means is it meant to diminish the
skills and years of experience of architects that do this for a living.
The fact is, if one can draw a rectangle, a circle, or a line, then one
has the skills to communicate an idea with clarity and thus kick off
the solution architecture process. A simple drawing expressing the
basic idea need not be a highly skilled or complicated effort at the
start; it can always be refined over time.

The process for developing a rapid solution architecture starts
with use cases and shown in Figure 9.1.

1. Ask: What problems will the solution solve? For whom?
2. At this step, consider the solution's potential to generate
   new business or help streamline business processes bal-

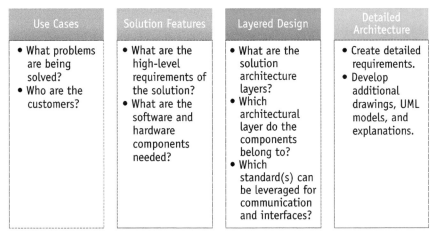

Use Cases	Solution Features	Layered Design	Detailed Architecture
• What problems are being solved? • Who are the customers?	• What are the high-level requirements of the solution? • What are the software and hardware components needed?	• What are the solution architecture layers? • Which architectural layer do the components belong to? • Which standard(s) can be leveraged for communication and interfaces?	• Create detailed requirements. • Develop additional drawings, UML models, and explanations.

Figure 9.1   Process for developing solution architectures.

anced against whether the costs and effort required are warranted.

3. If the answer to this critical business question is a reason that is not directly affecting an organization's profitability, competitiveness, or efficiency (e.g., "because it's cool"), the process should stop right there.

4. Apply common sense, think about business needs and opportunities, and recognize that technology solutions are aimed at meeting business needs.

5. Collaboration between the company's technology and business resources is a necessary part of the overall process.

The second part of the process is to consider necessary features and functions.

1. Ask: How will technology solve the problems identified? What are high-level solution requirements? What software functionality is required? What hardware and other systems will need to be integrated?

2. Features and functionality can be drawn as simple boxes, such as the example shown in Figure 9.2, with short text descriptions for what the box does.

Figure 9.2 Use simple boxes with text to describe features and functions.

3. The solution features are typically hard to get completely right on the first cut, but the features tend to grow over time as stakeholders point out aspects of the solution that are required to make it more complete.

Once the process of identifying high-level requirements and key functionality has been completed, the fun part begins in determining the architectural layer where each should reside.

1. If using a whiteboard, then at this point it will be easier to transfer the whiteboard feature boxes to a digital format, so that boxes can be dragged and moved to show each component's location within the architecture.

2. Because our focus is on IoT architectures, we know that there are three layers under consideration: the edge, the Cloud, and the enterprise.

3. Draw those layers and place the feature boxes in the correct layered location.

4. If there are existing standards or corporate architectural best practices, map those to the layers as well. (The standards mapping exercise can be complicated, but this chapter provides an example using the energy IoT reference architecture and IEEE 2030.5 mapping introduced in Chapter 8 to illustrate how the standards mapping can be included in the solution architecture drawing. Again, we chose IEEE 2030.5 because of its layered design for DER coordination with an IoT architecture in mind and the very clear relationship between the reference architecture and the standard.)

Completing these steps will yield a representative drawing of the solution architecture showing functions and features in their spatial location, a great start and a conversation piece to explain the idea and iterate on design. In practice, this first representative solution architecture drawing is often the one that gets revisited and used in presentations and conversations when introducing the idea to new stakeholders. It is simple.

However, it is also just the beginning of the architecture work. There will be many other forms of architectural drawings, typically determined by the architect's favorites: business process swim lane diagrams, sequence diagrams, UML, data models, stacked and block diagrams, and many other types of drawings; each may be necessary to clearly provide sufficient information for the development team to build the solution.

9.1.2   Benefits of Energy IoT Rapid Solution Architecture Methodology

The approach in this chapter provides numerous benefits. The Rapid Solution Architecture Methodology is designed to help facilitate brainstorming and conceptualization, organizing a solution idea into an IoT software pattern from the outset. Using a structured approach based on the energy IoT reference architecture as the common underlying architecture not only helps the innovation process, but also helps to keep an organization's solutions consistent with one another and allows for optimal reuse of code, data models, and services.

The process and examples provided in this chapter are intended to provide a methodology and toolset for people of all skill sets. The methodology depicted in Figure 9.3 is simple and fast. It can take as little as 30 minutes to scratch out a first draft of a conceptual solution. It is a use case-driven approach employed to solve specific business problems.

The use of the energy IoT reference architecture as the common reference architecture provides a consistent approach to leverage existing software and data patterns. The process allows for an independent path, to build on your own, or for easy collaboration with other stakeholders. The output is a simple visual architectural representation using the layered IoT architecture that is easy to communicate to all skill levels.

Figure 9.3 Reasons to use a rapid solution architecture approach.

No specialized tools are needed; PowerPoint or other common word processing and presentation applications can be used. If more sophisticated tools are readily available and the team has the skills to use them, then those architecture tools can be used as well. Examples of such more sophisticated tools include Microsoft Visio, Enterprise Architect, and ArchiMate. Draw.io is a freely available, simple-to-use toolset that provides the ability to create visual layers that can be turned on and off.

The architectural visualization output provides the high-level requirements and architectural elements. People of all skill sets can quickly understand where the architectural elements reside: at the edge, on a Cloud platform, or on premise at the enterprise. Interfaces, dependencies, and communication methods can be clearly included or implied by their location and functionality.

It is very important to recognize that this methodology is intended to be quick, efficient, and a transparent visual representation of a solution as a starting point. It is a conversation piece and a way to express the bigger idea. More architecture work will follow if the organization decides to move beyond a conceptual conversation to more detailed design.

9.2   Example Use Case

I chose this use case because it is currently an important and highly relevant topic within the electric power industry and a good example use case that the reader can use as is, expand upon, or use as a template for creating the reader's own DER integration use case. The remainder of this chapter will provide step-by-step instructions for developing a simple visual solution architecture that people of all skill levels can create and rapidly understand.

Behind-the-Meter DER Coordination Use Case
Interoperably coordinate and leverage residential behind-the-meter DER assets with utility operational systems to provide capacity and ancillary grid services and include an automated interconnect process for plug-and-play functionality.

9.2.1   Step 1: Create a Layered Template

The first step is to create a graphical template on which to lay your solution architecture, by marrying the standard you are using with the energy IoT reference architecture. One reason that we chose IEEE 2030.5 as the standard for this example is because of the direct mapping discussed previously.

In Step 1A, shown in Figure 9.4, begin by adding the three energy IoT reference architecture domains. For the purposes of this book and this example, a bottom-up architecture is one that begins at the OT edge and travels up through the Green Cloud to the energy business SaaS Systems domain. Drawing the energy IoT reference architecture like Figure 9.4 will help the reader to think from the edge up when the reader starts building solution architectures.

In Step 1B, shown in Figure 9.5, create the architectural layers for the standard using a simple line or box. For many standards, this could be a difficult process because there either is no real architecture to the standard or lining up the architecture with the energy IoT reference architecture may be challenging. More than likely, if a direct lineup of the reference architecture to one standard is not possible, you will need to use multiple standards to get your solution to work from the OT edge layer to the enterprise energy business systems layer. Regardless, whether using one standard or several standards, the same thought process works for rethinking how an IoT approach can help to simplify and scale for edge-focused solution architectures.

In Step 1C, shown in Figure 9.6, combine the templates from Steps 1A and 1B into a single template.

In Step 1D, save the template. This may seem obvious, but save the empty template because it is something you can use over and over to expand on existing solution architectures or to create

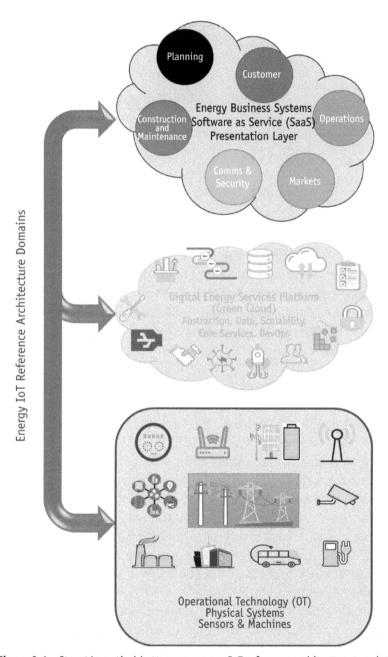

Figure 9.4 Step 1A: vertical bottom-up energy IoT reference architecture template.

new ones. The template in Figure 9.5 was created in Microsoft PowerPoint, so there are no special skills or software needed to create energy IoT reference architecture-compatible templates.

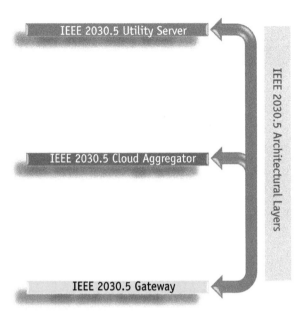

Figure 9.5   Step 1B: vertical bottom-up IEEE 2030.5 Protocol Standard template.

9.2.2   Step 2: Identify Energy IoT Reference Architecture Components for the Use Case

In Step 2A, starting from the edge, consider which architectural components or elements are required to support the use case. Simply circle those components on the template. For our use case, we have several components within the OT domain that will be necessary:

- *Sensors and measurement:* This will typically be the utility meter or some other measuring device to get load data and other current state information. (See Section 6.2 for more details on this architectural element.)

- *Gateways and local controllers:* If doing aggregation at the behind-the-meter premise, then gateways and controllers become the aggregation point. (See Section 6.3 for more details on this architectural element.)

- *DERs:* These are the behind-the-meter assets to coordinate with distributed generation (e.g., renewables, gensets), energy storage, and flexible loads. (See Section 6.4 for more details on this architectural element.)

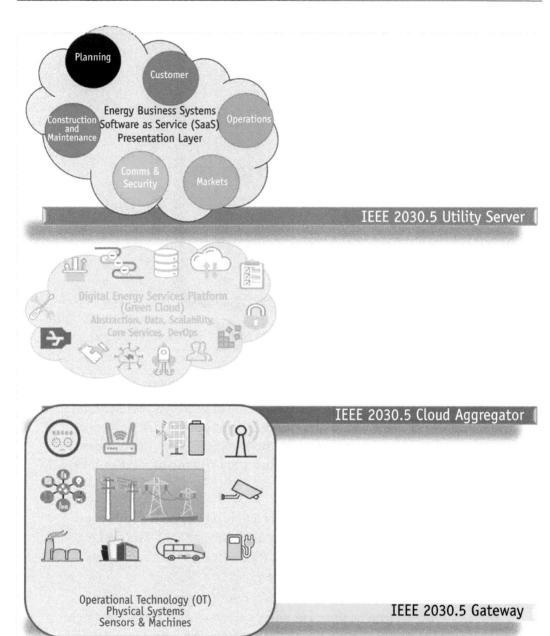

Figure 9.6 Step 1C: combined energy IoT reference architecture and IEEE 2030.5 Protocol Standard templates.

- *EV-charging infrastructure and EVs:* Special forms of DER can be engaged for managed grid services. (See Section 6.9 for more details on this architectural element.)

In Step 2B, moving to the Green Cloud platform, circle the energy IoT reference architecture components:

- *DevOps and software development:* This will provide developers with the environment for creating the use case solution architecture. (See Section 8.5 for more details on this architectural element.)

- *Microservices:* The Green Cloud's microservices for dynamic scaling of the solution architecture, containerization, and orchestration for deploying to the edge OT, and other microservice APIs to simplify communications with the OT and energy business SaaS system domains.

- *Event management and workflow automation:* Automated workflow processes and responses when different events occur (e.g., asset status reporting, grid services requests, communication losses). (See Section 8.6 for more details on this architectural element.)

- *Structured and unstructured data:* Processes and systems for managing, cleansing, and storing information (Section 8.7).

- *Security and identity management:* Role-based access control, encryption, and cybersecurity tools for securing communications and access. (See Section 8.8 for more details on this architectural element.)

- *Asset registry:* Tools for storing and providing information to other systems (e.g., asset type, capacity, location, asset owner, grid service capabilities, availability, supported protocols). (See Section 8.9 for more details on this architectural element.)

- *Digital twin agent:* Tools for simulation, communications, local data storage and processes, 3-D visualization, and more. (See Section 8.11 for more details on this architectural element.)

- *AI and optimization:* Tools to provide predictive analytics, optimization, troubleshooting, and simulation. (See Section 8.10 for more details on this architectural element.)

- *Aggregators and VPPs:* Tools to group assets. (See Section 8.12 for more details on this architectural element.)

• *Adapters:* Tools to translate different communication protocols for asset coordination.

In Step 2C, shown in Figure 9.7, the final step is to circle the
energy business SaaS systems:

• *Customer programs:* Utility system to include customers in
grid services programs. (See Section 7.2.1 for more details on
this architectural element.)

• *Interconnect:* Utility system to connect grid-connected customer systems for situational awareness and customer-pro-
vided grid services. (See Section 7.2.3 for more details on this
architectural element.)

• *DER management:* Utility system to visualize and coordinate
customer-owned DER for grid operations. (See Section 7.3.3
for more details on this architectural element.)

The entire Step 2 process may take as little as 5 minutes, and
when collaborating with colleagues, the number of architectural
components can easily be changed by adding or removing circles.

One interesting observation that the reader may have recognized is the number of architectural components circled in the
Green Cloud. These reusable tools are the power of a digital platform and the ability to abstract and simplify IoT communications
and data.

9.2.3   Step 3: Add Necessary Features to Support the Use Case

The third step is the part of this process that will typically take the
most time and will also likely become more refined as collaboration with others takes place. Step 2 identified the architectural components. Step 3 is more focused on what the solution architecture
will do. We call these features because they may be architectural
components, subcomponents, processes, services, or functionality.
They are added to help others understand the intent of the solution
architecture.

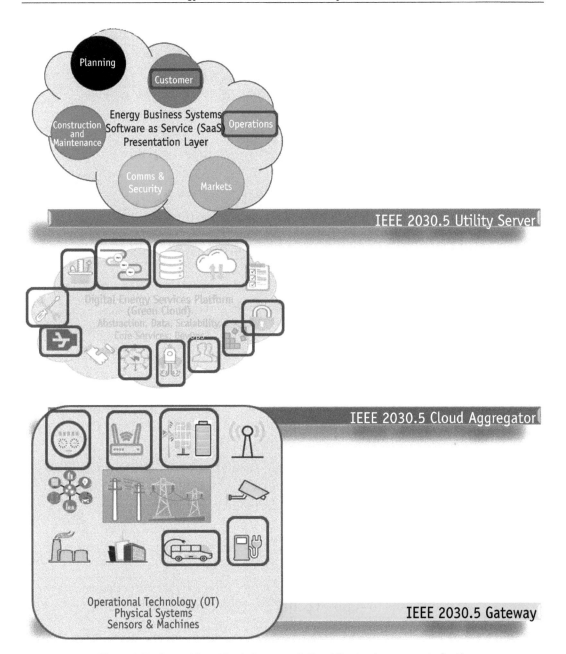

Figure 9.7   Steps 2A to 2C: circle energy IoT architectural components for the use case.

In Step 3A, again, starting at the bottom with the energy IoT reference architecture OT domain, consider the features needed to support your use case.

- *Gateway installation application:* Application used by installer to integrate with DER and register with IEEE Cloud aggregator system.

- *Gateway device and protocol adapters:* IEEE 2030.4 gateway with adapters for communicating with on-premise behind-the-meter DERs.

- *DER discovery:* Feature to discover added or removed behind-the-meter DERs.

- *Distributed generation:* Power-generating assets.

- *Flexible loads:* Loads that can be managed by the system.

- *Energy storage:* Batteries.

- *EV infrastructure:* Coordinated EV-charging systems.

- *EVs:* Coordinated EVs.

- *Metering:* Sensors that provide energy usage information.

In Step 3B, add use case features in Green Cloud to support Cloud aggregator systems.

- *Customer program participants:* The mapping of the customer program participants in the utility system to the behind-the-meter DERs participating in those programs.

- *DER capacity, capabilities, and availability:* Sizes, grid service support capabilities, and availability behind-the-meter DER information for visibility and dispatch algorithms.

- *Interconnect automated workflow:* As DER are added/removed from systems, this feature collects the information and makes it available to the IEEE 2030.5 server.

- *Customer participation opt-out data:* Collected data on customer opt-outs to support better prediction of future behavior.

- *Event performance data:* How well different OT systems performed during dispatch events.

- *DER groupings (VPP):* Grouping of individual and aggregated behind-the-meter DERs.

- *DER asset registry locations and demographics:* Geospatial and grid locational data for behind-the-meter DERs and IEEE 2030.5 gateways.

- *Customer energy performance portal:* Portal for the customers to observe the performance of their system (cost, energy usage).

- *Home energy management system:* System for customers to manage how their energy is used.

- *Customer behavior prediction:* Tool to help predict accurate grid services performance for behind-the-meter DER dispatch events based on previous customer opt-out behavior.

- *Event performance prediction:* Tool to help predict accurate grid services' performance for behind-the-meter DER dispatch events based on previous systems' performance.

- *Dispatch scheduling:* Dispatch schedules coordinating between the utility server and the gateway or behind-the-meter OT assets.

In Step 3C, as depicted in Figure 9.8, add use case features for the energy business SaaS systems:

- *DER visibility and situational awareness:* Near-real-time information of DER capacities and availability.

- *DER services event management:* Tools for behind-the-meter DER events, such as alarms, notifications, and status information.

- *Power flow modeling and simulation:* Support power flow with behind-the-meter DER information for more accurate modeling and prediction capabilities.

- *Customer program participants:* Sign up and maintain a database of participating customers in utility customer programs.

- *Operational and financial optimization:* Support algorithms for analytics to support optimizations that include behind-the-meter DER capacities and availability.

- *Interconnect data:* Data for a utility connectivity model to support power flow modeling and simulation.

In Step 3D, add any additional information that can help others to understand the use case's system design. For instance, adding information such as how the system will communicate

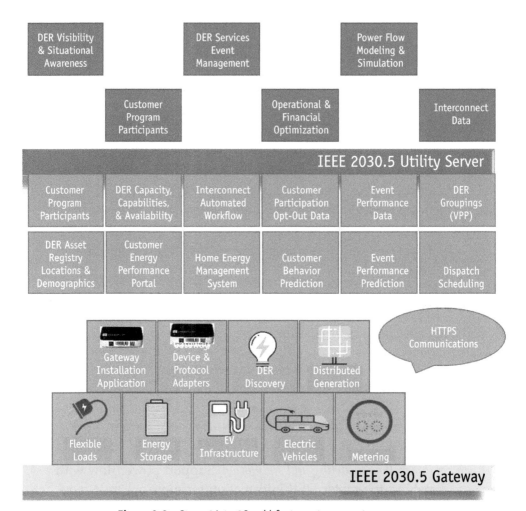

Figure 9.8   Steps 3A to 3C: add features to support use case.

(RESTful web services, pub/sub, cellular wireless, public internet, and telemetry) is helpful information that can be used to clarify your ideas. More sophisticated drawing applications such as Draw.io use visibility layers that can be turned on or off to help from making drawings overly messy.

Step 3 is the hardest part of the overall solution architecture development process, but an initial cut at it could take as little as 30 minutes. Collaboration with other colleagues is an excellent use of time to talk through the different features that the use case will need to support. Fine-tuning of the features will continue throughout the architecture development and even into development and upgrade discussions.

9.2.4 Step 4: Collaborate with Others

The solution architecture is an important tool for communicating your ideas graphically with others. Test your idea by talking through the architecture with colleagues. Listen to whether they understand the use case and how the architecture supports it. More than likely, your colleagues will not only find it interesting and even exciting, but they will also want to help you in refining the drawing and the way that it is communicated to other stakeholders.

9.2.5 Step 5 and Beyond: Create Supporting Architectural Drawings

The solution architecture methodology described in this book is just the beginning. It is intended to get people's brains working and to get them on the same page with the big picture of what it is. However, additional architectural drawings are needed to dive into the details to describe how it works. Business process diagrams, communications paths, enterprise system architectures, system interfaces, sequence diagrams, data models, and other architectural representations of the solution architecture will help in the development and future enhancements of the system.

9.3 Real-Life Examples of an Energy IoT Approach in Australia

Australia's path to decarbonization has been on a pace faster than anywhere else on the planet. The Australian clean energy market is estimated to be at least 3 years ahead of the U.S. market in integrating solar PV, batteries, and managing flexible loads.

Horizon Power (HP) is a small utility in Western Australia serving a very large and very remote territory in the western Outback. HP supplies power to remote communities in a service territory of 888,000 square miles (2.3 million km^2), which is equivalent to about 25% of the entire United States; imagine a service territory about the size of four Texas grids. It is an enormous area. Yet HP serves only about 110,000 customers, giving HP the dubious honor of serving the least number of customers per square mile on Earth.

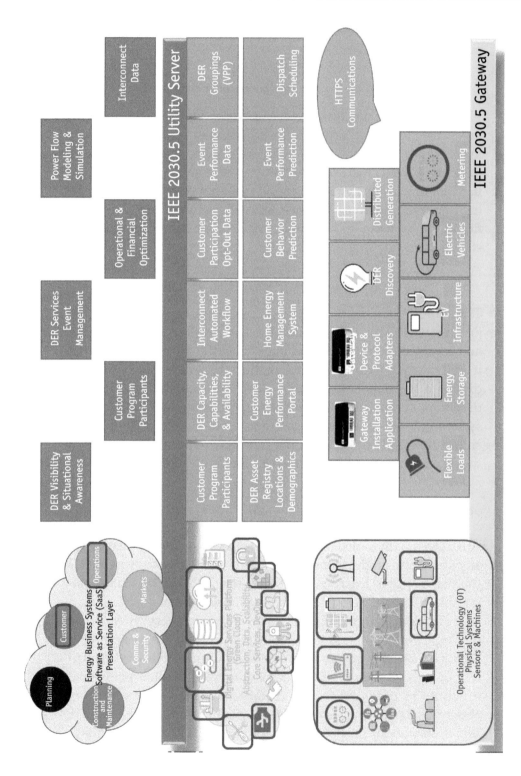

Figure 9.9 Draft solution architecture complete!

Due to the vast distances and small customer base, there is no regional transmission system to provide high-voltage bulk power services, so power supply to these remote communities is challenging. Figure 9.10 illustrates HP's territory and provides an aerial photo of the Onslow microgrid facility. The solution is that each remote community has its own separate, always-islanded microgrid that must generate, transmit, and distribute power to customers. HP manages 38 microgrids across Western Australia and is headquartered in Perth.

As one might expect, the primary sources of power for these remote microgrids are diesel generators, which are not environmentally friendly and emit GHG continuously when in operation, and power is needed by HP customers 24/7. In addition, supplying diesel-based power is expensive because diesel fuel is a commodity with fluctuating market-driven prices and the fuel must be transported large distances over water and land.

In an effort to be more cost-effective and to reduce GHG emissions, HP began an IEEE 2030.5-based DER integration initiative in the port city of Onslow in the Pilbara region, serving power to about 900 residents. The unique engineering, communications, cost, dependence on fossil fuel power generation, and distance

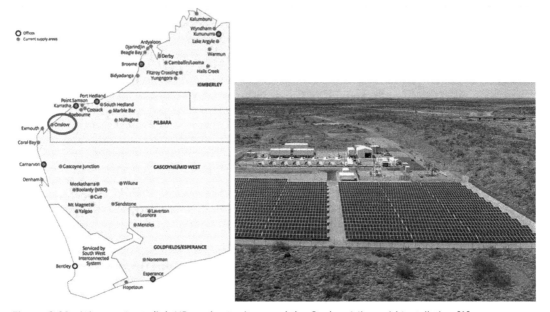

Figure 9.10   Western Australia's HP service territory and the Onslow Microgrid Installation [1].

challenges and HP's ambition to integrate DER to address all of those challenges have become one of the most referenced clean energy case studies in the world.

9.3.1 The HP DER Integration Experiment

With a small grant from the government, HP created a program to increase customer solar PV hosting capacity. Integration of behind-the-meter customer PV was a much different power delivery model, and HP executives, engineers, and project managers thoughtfully recognized the need for standardized, message-based communications. They chose IEEE 2030.5 as the primary protocol for communications. After conducting a tender process, they selected a commercial DERMS and a secure commercial gateway to the customer site to aggregate and coordinate with the DER assets.

The Onslow experiment was a smashing success and has become an often-referenced case study for integrating both in-front and behind-the-meter DER. The secure gateway device, called a Droplet and provided by the commercial software company SwitchDin, is a small and inexpensive industrial computer that communicates via the IEEE 2030.5 protocol and translates the protocol to several different DER communication standards (e.g., Modbus, DNP3, OPC-UA, and IEC 61850). SwitchDin's Droplet is quite flexible and can support a variety of DER use cases that coordinate and manage solar PV generation, batteries, flexible loads (primarily air conditioning compressors), and EV-charging infrastructure. Droplets were installed at customer locations and were also installed at HP's Onslow substation to manage a grid-scale PV array and battery. At HP's Perth operations center, the PXiSE DERMS system coordinates with DER aggregation services and the microgrid Droplets. The PXiSE DERMS and Droplet combination provides HP grid operators with real-time data from the grid-scale DER assets and aggregate data from the Droplets managing over 250 customer DER systems. The PXiSE system provides HP operators with situational awareness for the entire microgrid system using 4G LTE cellular communications. The Onslow case study utilized the entire IEEE 2030.5 IoT technology stack and has performed wonderfully since 2019.

In June 2021, HP performed a landmark, historic test at Onslow using only the available renewable energy assets and battery, a 100% hydrocarbon-free microgrid. The system operated flaw-lessly without fossil fuel generation for 80 minutes. The test used the 1-MWh grid-scale battery and 1-MW solar, along with its cus-tomers' solar and flexible load assets. This system has now become the poster child for grid decarbonization and how to properly inte-grate with utility and customer-owned DER. Much can be learned from that amazing little utility in Australia. Based on its lessons learned, HP is now replicating the success of the Onslow project and implementing clean energy solutions at numerous other mi-crogrid locations in its territory.

9.4   Conclusion

This solution architecture methodology provides a simple way for people of all skill levels to develop and understand solutions and use cases. Not every good idea comes from skilled architects or technical stakeholders, and this methodology does not require spe-cial skills or tools, just good ideas. The methodology and resulting architectural drawing are not intended to be the only architectural representation of the use case solution, and numerous other archi-tectural renderings are recommended. However, it is a good start-ing point to expand on an idea and grow it into a solution.

Reference

[1]   PXiSE Energy Solutions, "Horizon Power and PXiSE Energy Solutions Suc-cessfully Operate World's First 100% Solar-Plus-Storage-Only Community Grid,"   https://pxise.com/horizon-power-and-pxise-energy-solutions-suc-cessfully-operate-worlds-first-100-solar-plus-storage-only-community-grid/.

10

PNNL's Grid Architecture

The PNNL has developed a system architecture framework and advanced concepts for recognizing the changing patterns of today's electric power grid. The multiyear study [1] and set of work products provide grid architects with tools to help them manage the complexity of today's power grid. Led by the DoE's chief architect, Dr. Jeffrey Taft, the study combines general control theory concepts with network theory and system architecture methodologies, as depicted in Figure 10.1.

PNNL describes the grid as a system with ultralarge-scale complexity and the grid architecture framework was designed to help manage that complexity.

10.1 Laminar Decomposition

PNNL recognizes that the grid is evolving from a centralized, one-way power flow model of the twentieth century to a distributed, two-way power flow of the twenty-first century. With that profound challenge facing utilities, they introduced the concept of laminar decomposition, which is aligned with the hierarchical building blocks presented in the energy IoT reference architecture as the same fundamental grid design principle.

Laminar decomposition is the idea of designing the grid in a layered approach, with each grid structure able to manage itself

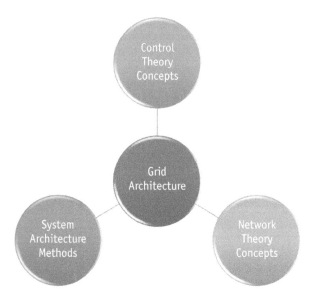

Figure 10.1   PNNL's grid architecture combines control theory, network theory, and system architecture methods.

and interoperate with other grid structures. The laminar decomposition terminology was the realization that you could think about grid systems in the same way that you think about a sheet of plywood with different layers bound together to create a strong yet flexible construction material.

Using laminar decomposition concepts, the resulting overall grid architecture becomes one that coordinates the different grid structure building blocks through layered coordination interfaces to create hierarchical relationships that are more resilient and extensible. Grid components become black boxes that combine within grid structures to create configurable grid systems as shown in Figure 10.2.

This relationship is the same concept of the neural grid described in Chapter 5 as well as the early discussion on thinking of everything as a microgrid. The grid architecture framework is focused on the middle piece, the structures and networks of structures that include regulatory structures, market structures, control structures, infrastructures, digital structures, and a coordination framework to integrate them. As PNNL sets out to design the grid from the bottom-up, with self-awareness, semi-autonomy, self-optimization, and interfaces to support larger grid structures,

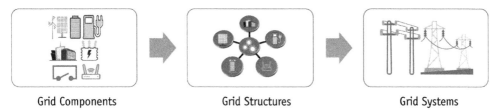

Grid Components   Grid Structures   Grid Systems

Figure 10.2   Relationship of grid systems to grid structures and grid components.

tomorrow's grid systems have great potential to be more resilient and, in time, they may even be able to reconfigure themselves to create greater efficiencies and adapt to both local and other system changes. The energy IoT reference architecture described in this book aligns very cleanly with PNNL's grid architecture framework to create a bottom-up hierarchical and adaptable grid.

10.2   Grid Architecture Framework Qualities and Properties

One of the foundational aspects of PNNL's framework is the concept of system qualities and properties.

- *System qualities:* PNNL describes these as the internal characteristics of the final system that can be expressed either qualitatively or quantitatively as the internal workings of the system (e.g., security and reliability).
- *System properties:* These external characteristics are recognized by the types of capabilities that they support (e.g., situational awareness and scalability).

Although both system qualities and system properties are inherently important, system properties enable the system qualities to be manifested, so a mapping relationship exists between the two aspects of the framework, as described in Figure 10.3. This mapping can be extremely complex, but PNNL developed a tool to help visualize those relationships.

In Figure 10.3, PNNL provides an example of its own mapping technique. In practice, a team of stakeholders develops the key components and structures, key properties, and system qualities boxes with the characteristics for each. PNNL recommends

Key Components and Structures   Key Properties   **System Qualities**

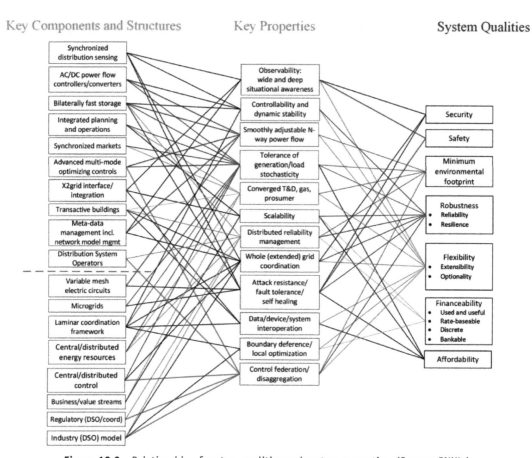

Figure 10.3 Relationship of system qualities and system properties. (Source: PNNL.)

starting from the middle column of key properties and then drawing lines to either side to indicate where relationships exist. This
can be a taxing exercise, but it underscores PNNL's assertion of the
ultralarge-scale complex nature of the electric power grid.

10.3   Other PNNL Grid Architecture Framework Features

PNNL has defined several different grid architecture structures involved in creating complete grid structures and systems, as shown
in Figure 10.4.

    PNNL accomplished groundbreaking work to help to capture
the numerous grid interfaces, components, stakeholder relation-

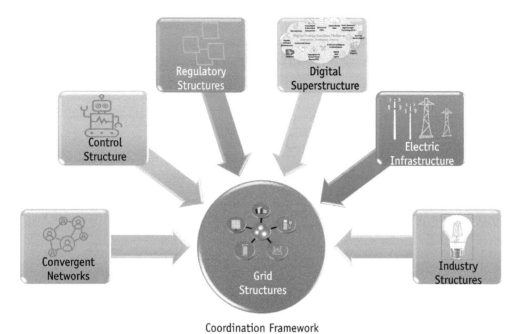

Coordination Framework

Figure 10.4   Grid architecture structures. (Source: PNNL.)

ships, and structures, but also a framework and methodology to visualize and manage complexity.

In addition to all the work cited by PNNL, they also created mathematical algorithms and methodologies to evaluate the complexity of system architectures using mapping, graph theory, and linear space/matrix methods.

PNNL offers valuable free resources through a dedicated landing page and numerous academic publications [1], which are well worth the time and effort to browse.

10.4   Conclusion

Inarguably, the grid is an ultralarge-scale complex system. PNNL has a library of free reference materials to help manage that complexity. Their grid architecture framework provides many definitions and a high-level understanding of layered (laminar) architectures, the numerous interacting structures, and several tools to help to visualize and quantify the complexity.

The energy IoT reference architecture, developed with the same concept of laminar decomposition, offers a specific three-layered IoT approach to abstracting and simplifying some of the enormous complexity illustrated in PNNL's grid architecture framework and toolsets.

Reference

[1] PNNL Grid Architecture, https://gridarchitecture.pnnl.gov/, Dr. Jeffrey Taft and others, 2022.

11

The Path to Decarbonization Requires Integrated DER

Utilities, third parties, and customers are installing DER assets in record numbers with no slowdown in sight. Most grid-connected DER assets are intelligent clean energy devices and can be engaged to support larger grid ecosystems and help them to become cleaner, more decarbonized systems of systems. Put simply, decarbonization is rapidly becoming synonymous with integrated DER/grid utility systems, most prominently with utility-scale wind and solar generation, systems that provide clean energy to substitute for fossil fuel energy.

Decarbonization = Integrated DER

DER can best be categorized in three groups: distributed generation (DG), energy storage, and controllable loads (sometimes referred to as DR assets or flexible loads). Table 11.1 provides some examples of different DERs.

Although energy efficiency is the first step to take in reducing energy consumption and beginning the path to decarbonization, energy efficiency is not a DER. DERs are distributed assets that produce, consume, or store electricity and can be controlled or coordinated with other parts of the ecosystem.

Special forms of DERs (EVs, microgrids, and energy storage devices) already provide a wide variety of grid services (e.g.,

Table 11.1
Common DER Types

DER Type	Examples	Comment
DG: renewables	Wind turbines, solar PV	Behind-the-meter customer-owned or in-front-of-the-meter utility grid-scale
DG: gensets	Third-party or customer-owned hydrocarbon generators	Used by IPP, customer cogeneration, or by customers normally for backup power
Controllable (flexible) loads	Heat pumps, chillers, CHP, pumps, water heaters, appliances, industrial processes, building management systems, other smart loads	Used for load flexibility and DR applications
Special DER: energy storage	Chemical and flow batteries, flywheels, pumped storage, compressed air	Devices that can store, consume, and deliver energy
Other special DERs	Electric transportation infrastructure and vehicles, microgrids	DERs that are mobile or that can island and support multiple use cases and grid services

frequency regulation, voltage/VAR, peak shaving, peak shifting). However, these particular distributed assets can be problematic for utilities to accommodate, as utilities have little experience integrating or coordinating such assets. That said, DER is evolving, changing, and creating new capabilities and services, so that utilities, even smaller utilities, will be better equipped to integrate DER in the near term.

11.1 Utilities' Path to Decarbonization

The falling cost, maturing technology, and zero to low carbon nature of DERs combine to make grid integration of renewables, flexible loads, EVs, and energy storage arguably the only rational path to decarbonized electricity. This section explores the path towards utility DER integration and shows why the utility's role in DERs may be considered as the cornerstone to continued reliability, power quality, resilience, and decarbonization success.

As the economic benefits of renewable energy, technology acceleration, and a social mandate for clean energy propel humans into a world that was not faced in the previous century, transportation and grid decarbonization are accelerating rapidly. Tesla is now the most valuable car manufacturer by market capitalization over long-standing companies such as Toyota, GM, Ford, and Volkswagen. The electric power industry is transitioning in both transmission and distribution operation and processes such as decarbonization are becoming a requirement to address carbon reduction and to remain relevant in the energy industry. Utilities can now be very strategic and pragmatic and make the choice of whether they want to play offense or defense as renewable energy and other DERs enter their networks.

As utilities ponder the impact of DER on their electricity networks, most recognize that there are both challenges and opportunities. There are so many questions demanding answers. Is greater DER penetration a threat to their business? Or is it an opportunity to provide new services and generate additional revenue? Is it both? How will they manage through this change? How does the organization need to change to be successful? Does the utility have the right foundations, such as architecture and technologies, to scale to support the large numbers that they are expecting? These are the questions being asked as utilities recognize that while facing change, they lack monopoly control over what customers and third parties will do behind the meter. The energy transition is happening much faster than most expected and continues to accelerate annually.

The grid is transforming from a centralized, fossil-fueled bulk power, system-centric, siloed architecture to a decentralized, decarbonized, event-driven, technology-rich, data-centric, energy IoT architecture. Understanding that pattern implied in the above questions can help utilities to plan, deliberately and analytically, for their future, so they may invest in no regrets internal activities and grid-modernization projects.

11.2   The Pattern for Utility Decarbonization

As utilities face this exciting transformation, it is sometimes a daunting task to identify a practical approach to planning and preparing the organization and technology solutions to recognize where the utility is headed and create the path to get there. Figure 11.1 provides a visual representation of the utility decarbonization planning process. The remainder of this chapter will focus on a step-by-step approach to such a planning exercise.

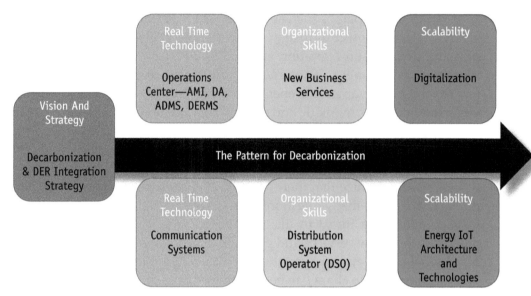

Figure 11.1   Utility pattern for decarbonization planning.

11.2.1   Step 1: Vision and Strategy

DERs are entering our networks and we do not know what they are, where they are, or how they affect our systems. We really need a plan.

The first critical step for an organization is to ask the crucial question of whether to play offense (i.e., active) or defense (i.e., passive). Many utilities have successfully played the defense strategy in the past: seeing change as a threat to the current business, waiting for things to settle down, but this time is different for the following reasons:

- A growing need for electricity (e.g., electrify everything).
- The electric power industry is becoming more democratized as new actors enter the industry and leverage new technologies and business services to disrupt slower moving utility companies (e.g., "Uberization" of taxis, car rental companies).
- First movers are already moving.
- Even more new actors seek to cash in on a massive industry with predictable growth.

The introduction and increasing penetration of DER into utility networks affect planning and operations, with impacts radiating out across the entire utility organization. Successful integration of DER into the utility business looks different to different departments. Defining success at the outset begins with a broad view, with planning at all levels within the utility.

As utilities become more comfortable with managing DERs, they also grow more confident and become more optimistic, able to see decarbonization as a real opportunity and to align their strategy with a more offensive posture. Some utilities now encourage customer DER adoption by increasing distribution hosting capacities and sometimes may even subsidize customer DER investments through proactive customer programs (e.g., Horizon Power in Western Australia). Success depends on a clear-eyed view of reality and then embracing creativity to craft new business services, models, and offerings to keep the utility relevant and profitable in the decarbonized future state.

During this first critical planning step, these questions must be answered:

- Are decarbonization and DER adoption a threat or an opportunity? Or both? Should they play offense or defense?

- Who are the stakeholders? External? Internal? What does success look like for each?

- What DER visibility and situational awareness are required? At what granularity? Does the utility need visibility for each endpoint, or are aggregated integration and reporting more optimal? If aggregated, at what level within the network does the utility need an aggregation? Building? Grid segment? Feeder? Substation? Other?

- What is the strategy to integrate decarbonization and DER adoption into planning, operations, and market systems? How much does the utility do itself, and how much should they rely on others? Can aggregators provide the proper data for forecasting DER impact on utility systems, or is higher granularity needed to ensure that DER does not inadvertently and unpredictably affect your production systems?

The utility can perform much of this analysis on their own and should include representation from across the entire utility company to gain perspectives and suggestions from everyone affected. Consulting companies can deliver paid support as needed to guide, manage processes, and create supporting documentation. The most successful path for determining an overall vision and strategy will start with executive sponsorship and a clear strategic vision, followed by internal workshops and inclusive brainstorming sessions. The full commitment of leadership and sound communication are critical to ensure buy-in from across the organization, so vital to sustained commitment and to generate excitement for great teamwork and innovation.

11.2.2   Step 2: Real-Time Technology

We may have a technology problem. Are our communications and operational systems adequate for all the future DERs that may be entering our networks?

Next on the checklist is recognizing that operational and communication systems may be insufficient to coordinate with large numbers of DERs. Depending on the granularity of DER reporting decided in Step 1, communication systems may need to support much larger numbers of assets than those for which they were designed. From an operations and cost perspective, what are the most efficient communications mechanisms for connecting and coordinating with large numbers of physically distributed assets? Operational systems must be reevaluated to consider how to interact with many small DER systems and to orchestrate them to perform as a highly functional grid system. All forms of DERs, generation, energy storage, and demand-side management assets are included in this coordination.

This leads to several questions that must be asked:

- Based on the granularity that we require from Step 1, what are our choices for communications? Is 5G ready and can it support communications across our territory? Is 4G/LTE capable of providing us with the proper granularity? How can we use our mesh networks, if at all, to coordinate with these new DER assets? Is fiber feasible or just too expensive to consider? With all these different options, which best serve us, our customers, and our partners?

- What is the future for SCADA-based communications? Will IP-based messaging protocols and information models replace point-to-point telemetry solutions? Will it be a hybrid of SCADA and these new approaches? What should the utility do to prepare for these changes?

- What are the limitations of our existing operational systems? Can they support communication and coordination of thousands, hundreds of thousands, or more DERs?

- How can our EMS/DMS integrate customer-owned, third-party, and utility DER assets? How much effort does this require? What are our current vendor's roadmaps to support both in-front and behind-the-meter assets? Are the state estimation, power flow, and forecasting tools adequate to support large numbers of small DER assets?

- Are there new utility DER customer program and DER aggregator market participation opportunities?

- What are our data needs for our operational systems in a DER-rich environment? Is it different from what was collected previously? Are there gaps? Where does it come from? How much data should we expect to transmit, store, and process? Can existing systems support these additional data volumes? Can we get the data easily, or are there specific technologies or processes needed to gain access?

After considering these questions, it is likely that existing communication and operational systems will need to change. Existing legacy systems will be called upon to support utility needs during this transition, as penetration grows, but may not be the appropriate future system to support very large numbers of DERs. In addition to the utility's communication infrastructure, the EMS, DMS, distribution automation, and DERMS must be carefully evaluated to identify near-term and longer-term organizational needs.

11.2.3   Step 3: Organizational Skills

What is the optimal combination of customer products and services? How does the organization need to change to support new products and services?

The third step is an organizational evaluation to assess the future utility organization and what services and products will be needed to maximize value. Decarbonization drives such organizational changes placing a premium on creativity unprecedented in relatively conservative utility cultures. Even as electricity consumption grows, selling more power is an unlikely utility business outcome as increased customer-owned DER and energy efficiency initiatives steadily reduce utility electricity sales. Customer-facing programs, power quality guarantees, and customer DER management and maintenance all become more common discussions for new services. However, this short list of new service concepts merely scratches the surface, implying that many unchartered opportunities have yet to be conceived.

As decarbonization accelerates, the idea of standard operating procedures (doing things like we always have) becomes less effective. Novelty emerges everywhere. Should utilities even consider entry into new distribution markets, for example? Each such decision carries with it significant economic, organizational, and technological implications. Utilities that may be considering becoming a DSO that also manages distribution markets, for instance, beg several functional questions (i.e., distribution grid operator, distribution market operator, distribution grid owner, DER manager, DER owner?). Determining which roles are the best future fit for the utility will help the organization to identify new skills needed and better prepare and plan for the decarbonized grid.

Questions about organizational skills may include:

- What new business services support a DER-rich ecosystem, and who is responsible for building them? Who benefits from these services, and what are those benefits? What is needed to build these services? Which new service proof-of-concept projects should be piloted?
- Will there be new local distribution markets to allow DER owners to monetize their investments and leverage a more holistic and resilient network model? Who should operate

these markets and how do they operate? Can the utility take part? What rules are expected to be in place for participating in these new markets?

• Who is responsible for managing, coordinating, and orchestrating DER on the distribution networks?

• What is the optimal DSO organizational model for the utility? For what does the utility want to be responsible for? For what does the utility not want to be responsible?

Organizational change is one of the most challenging areas for any organization, but especially those that have operated the same way for many years, often in siloed departments. Sound preparation to align with the overall vision and decarbonization strategy will help to ease the pain of significant change. Regular communication with staff and other stakeholders on a plan's progress will provide a sense of inclusion and clarify changing roles.

11.2.4   Step 4: Scalability

Throughout this book, I discussed the challenges with today's top-down, centralized, system-centric, siloed architecture with poor prospects to scale to millions of distributed physical assets. Issues of scalability have been a long-term and continuous conversation in standards organizations. The good news is that alternative, elastic, event-driven, data-centric, hierarchical, and decentralized architectures based on enabling standards stand ready to fill gaps as they arise. Modern-day architectures and technologies were designed to take advantage of virtualization, containerization, and orchestration, all of which are needed to deploy intelligence where it is needed, while also keeping it up-to-date and stable.

Cloud-based digital platforms and the IoT will enable distributed intelligence, allowing centralized utility operations centers to coordinate at system levels while local intelligence performs direct control of assets. This leads towards a more automated, intelligent, self-sustaining, resilient, and, most critically, scalable electric power industry ecosystem. As discussed in Chapter 10, PNNL's Grid Architecture program named this type of hierarchical layered control laminar decomposition. True transformation is on the menu now for the utility industry. Large Cloud digital technology providers and technology companies are already engaged.

Key scalability questions include:

- What will digitalization look like? Who is leading in this area? With whom should the utility work?
- Is disruption right around the corner? Who will they be? Should the utility be working with them? Are they competitors or partners?
- Who is paving the way where energy IoT architecture is concerned? What are the components that need to be considered? With whom should the utility work?

The pathway to scalability to accommodate massive numbers of DER assets runs through digitalization and an energy IoT ecosystem. Vendor partners will collaborate with their utility clients to ensure their products leverage new technologies and architecture. If not, utilities must seek out replacement vendors that are on the pathway to scalability. Scale remains the biggest problem facing the electric power industry, but fortunately, utilities still have time to plan and build organizations and technologies that can manage and optimize operations, new business services, customer programs, and profitability.

11.3   Conclusion

Like it or not, decarbonization has arrived. All electric power stakeholders must embrace the changes that decarbonization brings, choosing to collaborate and plan for opportunity rather than ignore or deny the pain, loss, and risks that inevitably accompany such monumental change. These patterns have come into focus across the utility world. Utilities that choose to be deliberate, realistic, inclusive, and creative will best weather the storms of change and avoid panic. Paradoxically, changes will not happen overnight, but they will most certainly occur faster than we anticipate. A very deliberate planning process will help minimize mistakes and ensure continuity of energy services that are timely, appropriate, and socially responsible to utility customers.

12

The Road Forward

We are at a very historic point in time. Climate change is real, and the entire world is experiencing the catastrophic effects of heat, storms, drought, fires, glacial melt, and ocean rise. Atmospheric carbon is the enemy, and our warriors will be policymakers, scientists, engineers, software developers, and technologists. The urgency in which we transform to a carbon-neutral global society has never been greater, not just in the electric power industry, but in all industries. However, decarbonization of the electric power industry has the biggest effect, the most impact. The industry is up for the challenge.

The IoT offers compelling technology capabilities that can create a future where massive numbers of distributed assets, distributed intelligence, and local autonomy transform the grid in very positive ways. An energy IoT–led transformation will unleash a new wave of innovation accompanied by some setbacks and the inevitable pain that comes with any change, especially one as big as this.

This book began with the three different business drivers (societal mandate for clean energy, economic advantages of renewables, and pace of technological change) that shape our changing electric power industry ecosystem and why the industry is ripe for true disruption. We followed that with architectural and technology challenges with the industry's more than 100-year-old model and how it will be unable to meet scalability and efficiency needs of a DER-rich ecosystem and why we need to rethink how we manage the grid. We introduced the energy IoT reference architecture theory that is highly scalable and event-driven and that leverages available Cloud technologies such as microservices, abstraction, virtualization, containers and orchestration, digital twins, data communications and storage mechanisms, use of existing IoT and energy protocol standards, and security. We discussed new energy service delivery models and potential use cases and new ideas that could result in plug-and-play simplicity for DER asset integration and coordination. We discussed alignment with other thought leaders and architectures such as the DoE PNNL's Grid Architecture methodology. We provided a step-by-step deep dive on how to turn theory into real-life practice using existing standards and the energy IoT reference architecture to rapidly create solution architectures that anyone who can draw circles, boxes, and arrows can build.

As a global energy transformation unfolds, we see evidence around every corner. For example, the U.S. government passed a bipartisan infrastructure bill that will flood the market with new clean energy innovations throughout the 2020s and beyond. The European Union (EU) penned the European Green Deal and proposed a law to ensure a climate-neutral EU by 2050. The EU had already committed to 50% carbon reduction by 2030 and now must redouble those efforts to realign after the Russian invasion of Ukraine challenged the status quo of imported natural gas. Whether driven by decarbonization policy, by the ever-declining price of renewables, by the pace of technological change, or by other grid forces in the news, all paths lead to the same conclusion. Change has arrived.

If we are to gain any valuable insight from the Covid-19 pandemic, we must recognize that our institutions are far more vulnerable to sudden systemic disruption than we had imagined. The catastrophic failure of the Texas electric grid in February 2021 provided more evidence that local resilience required more creative thinking to include leveraging DER to ensure that electric power is available during the worst of times. More than ever, we will need to invest in more local resiliency if we are to survive the upcoming potential shocks that we know are coming without catastrophic harm. We cannot know with certainty what a systemic crisis that disrupts our bulk electric power delivery might look like. Perhaps it will arrive as a cyberattack or a wave of hurricanes or more rampant wildfires, still more powerful and widespread. Whatever the cause, we must invest in a more distributed and resilient energy infrastructure that protects us all from historic disruption and other "black swans," whether they arrive as financial crises, terrorist attacks, horrific wars, more pandemics, or something altogether new and terrible.

The electric power industry is in historic times. A transformation is happening. The path forward is through digitalization and decarbonization and to electrify everything.

12.1   A Call to Action

Throughout this book, we have outlined a workable architecture that can enable a very rapid transformation to a distributed, robust, reliable, and lower-cost bidirectional power grid. The missing ingredient is aligning utilities, digital cloud providers, and vendors to step up and realize the vision. We require good science, engineering, applied technology, and leadership.

Nothing we have discussed is science fiction, and everything outlined in this book is possible now. All components shown in the energy IoT reference architecture currently exist or can be added. What we need is to educate the electric power industry stakeholders to move this future vision of the industry forward, however difficult that may be.

Cloud platforms and their supporting techniques stand ready to achieve this historic transformation expediently, safely, and inexpensively. Now globally available, these methods and technologies can offer the key digital services required to solve this puzzle. With this, a secure, compliant, and simple-to-use Green Cloud energy services layer will be created, where all core reusable services can be easily accessed, both securely and reliably from anywhere, whether from a Cloud or a local service provider.

We need to create a new type of service provider and an energy-specific DevOps environment for the electric power industry that allows us to simplify and accelerate how technologies connect and transact with energy grids nationally and globally. This new type of digital service provider will provide the connective tissue that coordinates with established energy assets and companies to facilitate new opportunities. To cater to the specific needs of the electric power industry, these services must be adaptable, low-cost, secure, and reliable. The provided services will enable new innovations from new actors, while simultaneously preserving security and reliability and lowering cost. Frankly, this is the only way to introduce the vital innovations that we need at scale, quickly enough to meet our shared aggressive global energy transformation goals.

Connecting solar, battery, and other DERs should, in short order, become as easy as plugging in any internet-connected electronic device (e.g., Alexa). The home energy management service will allow businesses and homeowners to connect to a nearby energy service provider as easily as we now connect to Amazon Video or Netflix in our homes. Based on the user's policy preferences configured during account setup, easy-to-use home and business energy management systems will discover newly added or removed assets and seamlessly connect them to the grid. These personalized policy preferences may include supported grid services or local market participation, among others.

We must adapt to regulatory, utility, creative, financial, and societal drivers by building a new type of digital energy Green Cloud service provider that puts us on a path to success and creates a secure, event-driven, distributed, transparent ecosystem. Such an ecosystem can enrich existing utility business models and foster the creation of new, innovative services in an ecosystem composed of innovators, businesses, and homeowners. We can open a door to new types of innovations so desperately needed to build a sustainable future that heals inadvertent damages from the old model. Another important outcome of this transformation is new well-paying job opportunities for both blue-collar and white-collar workers. Educators will find new areas of focus among an expanding spectrum of possible career opportunities that concentrate on our society's foundation, students and workers.

So how to make all of this happen?

12.2   The Big Picture Has to Work Together

The foundational energy IoT reference architecture drawing in Figure 12.1 is a simple representation of some very complex system-of-system relationships that were detailed throughout this book. The energy IoT reference architecture is abstract enough to easily represent the big picture, but also detailed enough to think about the individual components and have a good understanding of the role that they perform, the data that they produce, and how each element interoperates with other ecosystem components. The event-driven, data-centric, and loosely coupled architecture includes state-of-the-art security, including authentication, authorization, and access control roles that limit the probability of cascading events due to natural disruptions or a physical attack or cyberattack.

The three architectural domains (energy business systems, digital energy services platform, and OT) already exist in whole or in part. Together, these domains can support a transparent, fair, interactive ecosystem that is also data-centric, event-driven, secure, loosely coupled, and service-oriented. The ecosystem will enable distributed hierarchical command and control. Obviously these domains and the business models supported will all need to work together.

Founded in 2004, the GridWise Architecture Council, a team of industry leaders tasked to shape the guiding principles, or architecture, of a highly intelligent and interactive electric system, began working on the issue of interoperability. The NIST established the SGIP in 2009 with the same goal in mind. The SGIP was later merged with the Smart Electric Power Alliance (SEPA) and the original mission of the SGIP for greater interoperability of grid systems continues. Additionally, the DoE has continued its work on interoperability in the U.S. National Labs and through support of various standards-setting organizations (SSO). The ultimate end goal for interoperability becomes that of a plug-and-play ecosystem.

The fundamental need for interoperability is simple and pretty obvious:

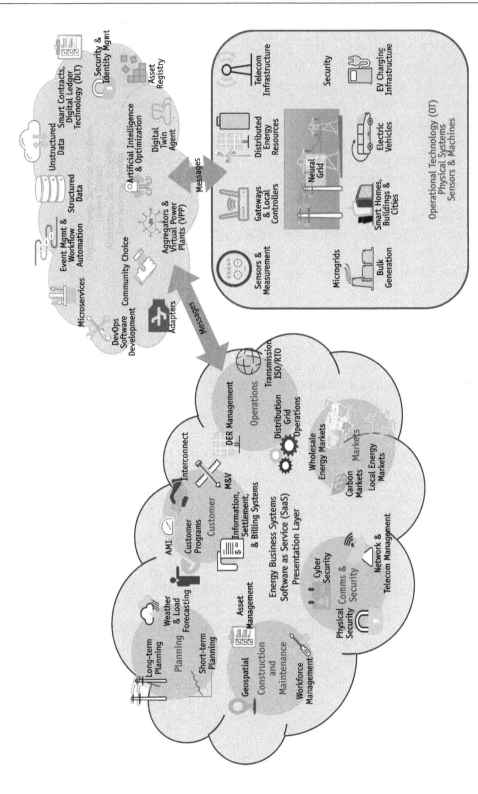

Figure 12.1 The energy IoT reference architecture.

- Lacking a common agreement on standards, connecting devices and systems will remain costly, time-consuming, and burdensome to maintain.
- Conversely, a common set of protocols, information models, and vocabulary for connecting devices and systems will reduce engineering time and cost.
- Because of legacy brown field systems and equipment, the large number of protocols used, and the large number of existing register-based communication with end devices, we will need adapters that can use a common communications methodology and translate them to the native protocol of the device for the foreseeable future.

Figure 12.2 describes the reinforcing relationship between interoperability, cost, and functionality. The more interoperable components are within an ecosystem, the less expensive it will be to create and maintain better systems that serve all stakeholders.

Today, processes and the preconditions are in place to reach a plug-and-play state. People and technology progress through stages, reaching consensus at each stage before arriving at an optimized end state.

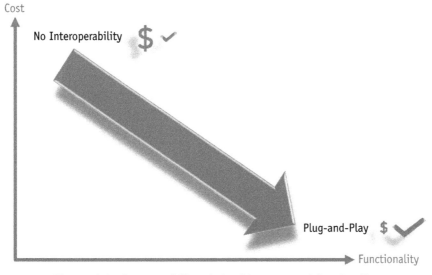

Figure 12.2 Interoperability relationship to cost and functionality.

Interoperability is incremental, starting small, with industry groups of people with shared mindsets that work together to clearly identify data requirements, use cases, and interfaces. Standards-setting organizations are consensus-based with clear procedural rules that require time, compromise, and a lot of work to reach agreement. The resulting standards are never static; they continue to evolve and over time this scales to match the needs of an industry. Historically, this approach has served us well and numerous standards are on just such a trajectory. When done correctly, standards can be extended and refined, gradually and methodically adding built-in functionality and lowering costs. The universal serial bus (USB) standard, now entering its fourth generation, stands out as a great example, providing dramatically faster charge speeds, as customer-demanded improvements and vendor collaboration on extended specifications led to greater functionality over time.

12.3   Leveraging DER to Provide Resilience

Resilience has become the new electricity service now sweeping the globe. Commercial vendors develop microgrids and other resilience services to entice utility customers away from their legacy electricity service provider. Events such as the Texas freeze in 2021, where 70% of the Electric Reliability Council of Texas (ERCOT) customers lost power, have spawned a resilience industry that invites vendors to poach the best customers from the utility. A majority of the conventional multinational utility vendors (e.g., Schneider Electric, Siemens, ABB, General Electric (GE)) have microgrid solutions in their portfolios. Newer technology companies such as Generac, Uplight, and Enchanted Rock actively engage with customers to win business away from utilities.

There is a growing interest in implementing microgrid solutions for commercial and industrial businesses. Long-term PPAs not only remove these important commercial and industrial customers from utility portfolios, but also offer them highly competitive pricing in conjunction with the potential for enhanced energy resilience. Increasingly, public power and investor-owned utilities risk losing their best customers if they cannot provide reliable, resilient power. Utility companies must learn how to integrate and manage DERs at scale and potentially offer their own microgrid solutions to compete with commercial microgrid providers. Thankfully, the stimulus packages making their way through Congress afford utilities with an amazing opportunity to collaborate with vendors and consulting companies to apply for grants to create these capabilities.

In summary, scaling EVs, PV, solar, and energy storage creates new challenges for conventional utilities, and the competitive playing field is getting crowded, as commercial companies ramp up to cherry-pick the best utility customers with cost-effective and versatile solar plus storage and competitive/resilient microgrids.

12.4   The Way Forward

The magnitude of transformational change and how to manage that change are hard to imagine. It does not help that the required activities for this much change are a departure from the way that the electric power industry has done things for over the past 100 years. We need a pragmatic approach that takes this very big idea and breaks it down into manageable tasks and milestones to support quick actions and periodic measurement.

One possible pathway to the future is a four-step approach that engages with one or more progressive utilities, technology companies, and vendor partners as depicted in Figure 12.3:

1. Align the partners.
2. Build the critical services.
3. Pilot the approach.
4. Implement at scale.

12.4.1   Align the Partners

This pathway to the future would need to organize a select group of executive sponsors that are technologically savvy to support innovation workshops taking place over a calendar quarter. This group of executive sponsors should consist of utilities, Cloud providers, industry vendors, government agencies, and regulators, all with the goal of devising a plan to apply the concepts of the energy IoT reference architecture to existing use cases and answer:

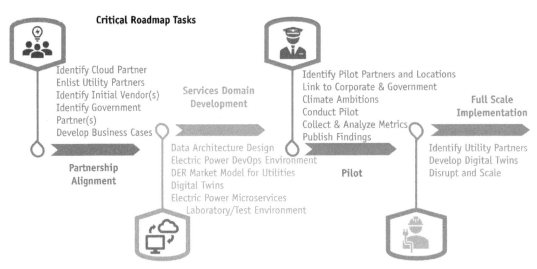

Critical Roadmap Tasks

Identify Cloud Partner
Enlist Utility Partners
Identify Initial Vendor(s)
Identify Government
Partner(s)
Develop Business Cases

Services Domain
Development

Identify Pilot Partners and Locations
Link to Corporate & Government
Climate Ambitions
Conduct Pilot
Collect & Analyze Metrics
Publish Findings

Full Scale
Implementation

**Partnership
Alignment**

Data Architecture Design
Electric Power DevOps Environment
DER Market Model for Utilities
Digital Twins
Electric Power Microservices
Laboratory/Test Environment

Pilot

Identify Utility Partners
Develop Digital Twins
Disrupt and Scale

Figure 12.3   Implementing the energy IoT conceptual architecture through utility,
technology, vendor, and industry partnerships.

- What current use cases would create the most benefit and greatest opportunity for success?
- Where are the best opportunities within the electric power industry for their implementation?
- What are the present execution capabilities, gaps, dependencies, and blind spots?
- Where can the industry test these proof-of-principle pilots safely?
- What is the execution roadmap?
- What are the investment priorities?

This group of companies, agencies, investors, and individuals will finance and work together to drive to a common future state energy IoT ecosystem. Additional funding may come through grants, sponsorships, and partner-in-kind contributions. Acting as an executive steering committee, this team will act as technical champions, help to connect with available experts and thought leaders, prioritize key issues and challenges, prioritize use cases, review progress and milestones, provide guidance, and remove barriers to success.

12.4.2   Build the Services

As stated previously, none of this is science fiction and most, if not all, of the technology already exists. Unfortunately, the generic Cloud platforms and IoT toolsets available are not specifically tuned for the electric power industry, requiring strong industry knowledge as well as DevOps software development experience and one-off solutions that support specific use cases or business models. In order to gain large-scale acceptance, the industry needs a tailored set of microservice components and tools that are packaged in a digital Cloud platform so that they are reusable, scalable, and secure. The special packaging needs to include common standards-based asset and information models with common vocabularies, DevOps tools for source code version control and team sharing, standard microservices for scaling and containerization, additional microservices for digital twin agents, adapters, and communications, data storage and data cleansing systems for big data, analytics and optimizers for greater efficiencies, and security and privacy tools to protect data and communications.

Some baseline assumptions and expectations to execute this book's proposed roadmap and package this energy-specific Cloud-based ecosystem include:

1. *Assemble the best and brightest in the electric power industry.* There are many passionate and dedicated architects, thought leaders, technology professionals, electric power operations, and market experts who can be engaged to coordinate this highly complex technological undertaking. There will be mistakes and failures. We must be patient, fail fast, and learn quickly from our mistakes. This is research and development requiring an iterative approach, the potential benefits of which will generate large returns in the long run.

2. *We will leverage digital Cloud technologies.* A technological revolution over the last 20 years has led to the widespread adoption of a standardized set of internet technologies that now enable Cloud computing at scale. A huge workforce now familiar with this vast set of technologies and techniques has to date focused more on commerce and entertainment and less on industrial-scale processes. Howev-

er, the rapid transformation from an ownership to a solution paradigm in the transportation sector (e.g., Uber and Tesla) is promising for a similar shift in the electric power industry. We must now apply the secure, flexible, distributed, and elastic on-demand computational capabilities of Cloud platforms to successfully achieve the global energy transformation.

3. *We will need a common data architecture.* Developing a data architecture will perhaps be the biggest challenge of all. Many different asset model standards exist, and this is same situation with structured and unstructured data: some are relational, some are in very large data sets, some come with privacy issues, some must be accessed by numerous services and systems, and most, if not all, will need to be secure. Assets will need to be registered with unique identification values within the ecosystem. Many relational associations will exist between data, use cases, companies, people, roles, markets, service providers, assets, locations, aggregation and points of coordination, and networks. Finally, to avoid confusion and take advantage of existing information models, profiles, and protocols, data will require a common vocabulary, perhaps based on IEC's CIM and 61850.

4. *Use what we already have.* Many legacy applications, built from a system-centric perspective rather than a data and event-driven perspective, provide the electric power industry with complex and expensive integration solutions that remain brittle and difficult to maintain, and yet they work. In some cases, these siloed systems were designed that way purposely, but they were all built using today's top-down, one-way power flow, centralized command and control architecture. Their entirety or parts of them can be converted to microservice APIs, reusable components that can evolve to support new use cases and capabilities. Major rework will be required to thoughtfully allow them to align with the energy IoT reference architecture and still operate in parallel as a componentized redesign is developed. The logic and algorithms can be harvested to create

componentized services-based APIs to apply best practices for digital Cloud development and maximize existing code reuse.

5. *We will design for interoperability.* The use of standards cannot be understated. We will use existing standards, information models, profiles, and communications protocols that will increase our ability to be interoperable with our own solutions and with others. Standards will need to evolve to support new functions and will need to be extended, but wherever possible, the use of standards will help support backwards-compatibility with legacy brown field assets and systems.

6. *We will select simple, but measurable use cases in the beginning.* We will select use cases with a strong chance of success, but the use cases that we select will also require the development of services and capabilities to support more complex use cases to follow.

We will identify appropriate laboratories to test our solutions and use cases.

Small and large-scale testing facilities will be needed. These could be provided by government or private industry. At present, the Idaho National Laboratory (INL) shown in Figure 12.4 has two such laboratories, Next Generation Wireless Test Bed and Electric Test Bed, which may be good candidates. Regardless, multiple locations are practical to test in different environmental conditions to accurately test hypotheses, use cases, and technologies at scale with real equipment.

Developing the Green Cloud digital services domain will require new microservices, a data architecture and common vocabulary, accurate and detailed asset models, and digital twin agents and adapters to support simulation and abstract communications and coordination with OT assets. Some microservices that could perform forecasts (building loads, EV loads, other asset loads, PV generation, wind generation, genset generation, weather, market prices, fuel usage, CO_2 emissions) and manage data access, edge and cloud-to-cloud communications, aggregation, and event management are obvious potential APIs to develop.

Figure 12.4   INL's National and Homeland Security test facilities are potential sites to test with real equipment.

12.4.3   Pilot the Approach

Once a suitable laboratory has been identified, the next step is to assemble a team of ambitious and passionate stakeholders to develop a pilot that tests initial use cases. UML modeling tools such as Enterprise Architect may be used to create unambiguous documentation for the stakeholders involved in the pilot project. A variety of interoperability scenarios, electric power industry standards, and communications pathways and protocols can be exercised to help vendors to align their products and services to interoperate with utility systems, Cloud microservices, and OT assets in an event-driven, data-centric, distributed architecture. Interoperability using a three-layered IoT architectural approach is the pilot's primary goal.

A good example of an initial pilot is the abstraction and integration of behind-the-meter home and small business DER, which requires a simple home energy management system (HEMS) that discovers newly installed or removed DERs and manages the DER assets based on control or market signals. A solution architecture example for this use case was provided in Chapter 9. Horizon

Power implemented this capability through a residential gateway product from a commercial provider (also discussed in Chapter 9).

This commercial gateway provides several different functions such as providing a universal protocol translator (adapters) to communicate with energy assets, integration with cloud aggregator services and utility operations systems (DERMS), aggregation of DER assets, common/abstracted communications to the edge, and also the customer's ability to manage his own DER assets. Cloud platform provider giants such as Apple, Google, and Amazon could provide similar gateway services with Homepod, Home Hub, and Alexa, respectively. A pilot such as this could test a number of different scenarios such as utility control of DER assets, coordination with other intelligent grid assets (intelligent electronic devices (IED)) for targeted local grid services, and perhaps even customer or utility participation in wholesale and distribution markets.

On completion of pilots, metrics must be fully captured, lessons learned, critical issues experienced, and success formulas devised and published before new use cases are piloted. Ideally, discoveries, methodologies, and publications will reside free of use in the public domain. There may even be an opportunity to provide educational workshops for utilities, vendors, regulators, educators, and other stakeholders.

12.4.4 Implement at Scale

A large collaborative industry pilot will pave the way for adoption with new opportunities for true innovation, greater resilience, and a clean energy future. The pilot must leverage modern digital Cloud technologies, demonstrate the power of a scalable ecosystem, support a highly distributed set of assets regardless of ownership, and operate in an efficient, elegant, and event-driven way.

Those involved with the energy IoT roadmap and pilot will have the first opportunity to implement their solutions at scale. Team members will work directly with partner utility personnel and advisors to thoughtfully identify affected processes and systems before the pilot solution can be deployed within their organization. As other systems and services migrate to an energy IoT architectural approach, the process will get easier. That said, even in the early days of integrating new energy IoT solutions, the amount of effort required will likely be less than what is needed to integrate multiple disparate systems in today's top-down, centralized, siloed architecture.

12.5   Will You Be Part of the Solution?

Many of the most passionate, progressive, most innovative, smartest, and bravest people and the most innovative companies in the world are included in the following list. We challenge leaders from these organizations to work to make the energy transformation a reality in this decade:

1. Utilities spanning the investor-owned, public power, and co-op sectors that are willing to solve their most difficult operational problems with the architecture and technologies envisioned in this book.

2. Tech companies like Google, Microsoft, and Amazon will need to work with the industry to develop the Green Cloud abstraction layer. The businesses with the most intelligent staff, best technology, and willingness to undertake risk while addressing big societal issues will be successful.

3. Small and large energy vendors such as Siemens, Schneider Electric, ABB, GE, SEL, and others will need to develop hardware and systems that conform to interoperate within this new ecosystem.

4. Universities will need to teach the energy IoT reference architecture concepts, establish laboratory environments to build and test solution architectures, and engage students to become the new leaders in creating interoperable clean energy solutions.

5. Government agencies such as the DoE, NIST, NASA, or IPCC will need to fund a panel of 15 to 20 hand-picked architects, data modelers, and semantic language experts (IEC CIM, 61850, IEEE 2030.5, OpenFMB, for example) for 1 year to expand on the energy IoT reference architecture and create common data models to enable this transformation.

6. Standards development organizations such as IEC, IIC, or IEEE will need to create a digital twin agent standard in 18 months for the electric power industry.

7. Public utility commissions will need to design policy and begin pilots to democratize distribution markets. This will inspire invention and the implementation of renewables and energy storage, while creating new opportunities for energy consumers and producers.

8. Pilot funding through government and philanthropical organization grants, private equity and vendor investments, and public-private partnerships.

9. Further, it is necessary that politicians recognize this monumental need and economic opportunity that generates a sustainable political will to address the societal drivers for this change. They must create practical policy that not only encourages transportation electrification and energy storage adoption, but must also offer grant money for ARPA and DoE to provide funding for both established and fledgling companies with innovative solutions that directly align with the energy IoT ecosystem. It is imperative that this transformation carries the same national importance of the space race of the 1960s.

This list is not by any means exclusive or the last word. Suffice to say that those who read this list will recognize themselves and step up to the challenge. The future of our planet depends on three objectives: digitalization, decarbonization, and electrify everything. Those in our generation willing to take on one of the great adventures for all time will rise to this moonshot moment to combat climate change, simplify and reduce the amount of time to add DERs and other clean energy assets, and advance our grid to become more resilient, sustainable, and reliable: a neural grid. We need people like you.

Reference

[1]   Idaho National Laboratories website, https://inl.gov/.

Appendix
Relevant Communication Protocols and Standards

A.1 IEEE 2030.5

See Chapter 8.

A.2 OpenFMB

OpenFMB is a NAESB standard and reference architecture. Like 2030.5, it is a profile of CIM and IEC 61850. It was started by Dr. Stuart Laval at Duke Energy, who initiated a "coalition of the willing" to work with vendors to create a fully pub/sub, use case-driven, IP-based information model and architecture. This low-latency model is ideal for managing utility-owned grid assets and for real-time microgrid operations. It someday may be a replacement for DNP3.

It is a very promising standard with an operational (and laboratory) deployment at Duke's Mt. Holly facility. It has great potential for running the internal operations of the microgrid. In 2021, OpenFMB introduced a certification program.

OpenFMB enables distributed intelligence and laminar architectures. It supports the layering of multiple functions and use cases for stacked benefits. Due to its use of unified semantics based on the CIM and IEC 61850, it enables the capability for advanced analytics. It supports plug-and-play interoperability with devices and systems through the use of adapters. It can provide local sub-second fast response services when centralized command and control are suboptimal for communicating in a timely manner.

OpenFMB enables local resilience when portions of the grid are segmented or islanded. It actively coordinates assets to operate within the local power system constraints. Security was designed into the standard from the beginning using TLS encryption techniques. It also supports several different pub/sub protocols including MQTT and DDS which can allow for very granular encryption and authorization at topic or message levels.

OpenFMB is fully open-source with freely available schema to download. The OpenFMB standard can be purchased from NAESB for $300. It is currently being updated with a modified information model and schema. An additional testing and certification standard is being finalized that will accompany the updated OpenFMB standard.

A.3   IEC 61850

IEC standards are used throughout the world, especially in Europe and Asia. However, their use within the United States has been very limited. IEC 61850 has a very rich information model for grid assets, including DER. IEC 61968/70 is commonly referred to as the CIM and was used with IEC 61850 as the starting point to create the IEEE 2030.5 and OpenFMB profiles. The information model may also be used as the underlying data model for the Green Cloud.

A.4   Open Automated Demand Response (OpenADR)

OpenADR was championed by the American Society of Heating, Refrigerating and Air-Conditioning Engineers (ASHRAE) building management stakeholders as part of the SGIP effort to help utilities communicate with building automation systems in a standardized way.

The DR functions available in OpenADR are completely duplicative of the functions available in IEEE 2030.5, but 2030.5 is much more powerful. That said, OpenADR has an established ecosystem within the building management community, where 2030.5 has not yet taken hold. Over time, the OpenADR standard may be deprecated by the utility and building management communities in favor of 2030.5. Until then, meeting DR requirements will probably require an OpenADR interface when communicating with building energy management systems (BEMS).

A.5   IEEE 1547

The IEEE 1547 Standard for Interconnecting Distributed Resources with Electric Power Systems is one of the most important standards for how DERs connect to the grid. Compliance with this standard is an absolute must for North America. Section 1254, "Interconnection Services of the U.S. Federal Energy Policy Act of 2005," established IEEE 1547 and future amendments as the national standard for DERs interconnecting within the United States.

IEEE 1547 is a series of standards with six other accompanying and complementary standards (Figure A.1) that provide implementation guidelines, security best practices, interconnection best practices, design and operation guidelines, and testing requirements. The 1547 series of standards are under the jurisdiction of IEEE's Standards Coordinating Committee 21 on Fuel Cells, Photovoltaics, Dispersed Generation, and Energy Storage (SCC 21).

The standard focuses on providing functional technical requirements for DER electrical interconnection with the grid. It does not describe the types of DER or define the types of technologies or communication standards used for DER. In other words, it is technology-neutral.

IEEE 1547 addresses the electrical interconnection of DER and their interoperability between electric power systems. This includes requirements for interconnection to both primary and secondary distribution voltages. The requirements provided are concentrated on safety, operation and performance, testing, and maintenance of the DER interconnection. It provides specific requirements for anti-islanding, voltage regulation and power functions, and ride-through functions. The standard also provides

Figure A.1   IEEE 1547 series of standards.

requirements for testing, commissioning, installation, and design. One important consideration with IEEE 1547 is that it was written for North American power systems, so the standard considered a 60-Hz frequency in its development.

IEEE 1547 is available for purchase at the ANSI Webstore and the standards in the series can be purchased individually or as a bundle.

A.6   Modbus and SunSpec Modbus

Modbus was originally developed by Modicon (now part of Schneider Electric) in 1979 for programmable logic controller (PLC) systems. The standard is frame and registry-based. In other words, it is a way to read and write bits (setpoints) on a piece of machinery. The transport mechanism can be serial or ethernet communications. Because it is a PLC standard, it is configured in a master/slave relationship with the PLC being the master and whatever equipment it is communicating with is the slave. So the PLC polls the equipment as the master and the slaves can respond, but no slave can initiate communications.

The SunSpec Modbus standard was one of the standards identified in IEEE 1547, which is the DER interconnect standard for all DERs connected to the grid. SunSpec developed a standard map for the Modbus Frame and Registry settings. A common XML model definition can be downloaded at https://sunspec.org/sunspec-information-model-specification/. There are many SunSpec Modbus products available, especially in the energy storage and solar PV communities. SunSpec Modbus compliance is self-certified by the vendor and vendors are not required to support all the functions.

A.7 OCPP and OSCP

The most popular standards for EV charging are the OCPP and the OSCP. These standards are led by the Open Charge Alliance and are rapidly evolving to support a large number of EV use cases, including vehicle-to-grid (V2G) support.

OCPP is a flexible, extensible, message-based protocol used to support interoperability between the EV, EVSE, and the charging point operator (CPO). The protocol can provide the status of the charging station and manages EV-charging transactions, different payment options, timestamping, EV configurations, metering values, driver and charging operator messages, authorization, charging reservations and schedules, internal load balancing, and other smart charging services. Security is implemented through the ISO 15118 TLS certificates. The Open Charge Alliance includes full documentation that includes the architecture and topology, specifications, schemas, certification profiles, test cases, and implementation guidelines. Use cases with numbered requirements and sequence diagrams are also well documented. OCPP has become the dominant EV-charging protocol used internationally. Certification services are available through several independent test labs globally. The open standard is freely-downloadable with no licensing or royalty fees.

OSCP is additional functionality for interoperability with the CPO and other entities, such as the utility or Charging as a Service (CaaS) provider, requiring capacity and availability information. The OSCP helps to predict the local available capacity to the CPO and on to a DSO to support charging optimization with larger energy ecosystems such as transportation microgrids that include PV solar, generators, energy storage, and other DER assets. The standard includes use cases and utilizes JSON/RESTful services.

A.8 IEEE 2030.7

Although IEEE 2030.7 is not a communications protocol, it has some very useful information in it and provides specifications for microgrid controllers. It provides a baseline for microgrid functionality. Microgrid functional requirements were developed using the following use cases:

1. Resilience:
    (a) Backup power during outages;
    (b) Black start from islanded mode;
    (c) Load management and prioritization: management of controllable loads.
2. Reliability:
    (a) Uninterrupted seamless service when entering island mode;
    (b) Improved SAIDI, SAIFI, and CAIFI;
    (c) Management of power quality such as frequency regulation and volt/VAR;
    (d) Reconnect and synchronization with the grid during transitions from islanded mode to grid-connected mode and vice versa;
    (e) Operate in islanded mode.
3. Economic optimization:
    (a) Optimization of available microgrid resources;

    (b) Optimization of grid incentives and market prices;

    (c) Load shifting to reduce grid-supplied power during peak power events.

4. Meet interconnection requirements for grid operator.

List of Acronyms

ADMS   Advanced distribution management systems

AEMO   Australian Energy Market Operator

AI   Artificial intelligence

AMI   Advanced metering infrastructure

API   Application programming interface

ARPA-E   Advanced Research Projects Agency-Energy

ASHRAE   American Society of Heating, Refrigerating and Air-Conditioning Engineers

AWS   Amazon Web Services platform

BTM   Behind-the-meter

CAISO   California Independent System Operator

CHP   Combined heat and power

CIM   Common information model

CIS   Customer information system

CO_2   Carbon dioxide

CPO   Charging point operator

CPUC   California Public Utilities Commission

CSIP   Common smart inverter profile

DER   Distributed energy resource

DERMS   Distributed energy resource management system

DGO   Distribution grid operator

DLT   Digital ledger technology

DMO   Distribution market operator

DMS   Distribution management systems

DNP3   Distributed Network Protocol 3

DNS   Domain name server

DNSP   Distribution network service provider

DoD   U.S. Department of Defense

DoE   U.S. Department of Energy

DR   Demand response

DSO   Distribution system operator

EaaS   Energy as a Service

ECC   Elliptic Curve Cryptography

EIA   U.S. Energy Information Agency

EMS   Energy management system

ERCOT   Electric Reliability Council of Texas

EV   Electric vehicle

EVSE   Electric vehicle supply equipment

FERC   Federal Energy Regulatory Commission

GFEMS   Generation facility management system

GHG   Greenhouse gas

GIS   Geographic Information System

GW/GWh   Gigawatt/gigawatt hour

HAN   Home area network

HEMS   Home energy management system

HTML   Hypertext Markup Language

HTTP(S)   Hypertext Transfer Protocol (Secure)

HVAC   Heating ventilation and air conditioning

ICE   Internal combustion engine

IEC   International Electrotechnical Commission

IED   Intelligent electronic devices

IEEE   Institute of Electrical and Electronics Engineers

IIoT   Industrial Internet of Things

IoT   Internet of Things

INL   Idaho National Laboratory

IP   Internet Protocol

IPCC   Intergovernmental Panel on Climate Change

IPP   Independent power producer

ISO   Independent system operator

KW/KWh   Kilowatt/kilowatt-hour

LDA   Local distribution area

LMP   Locational marginal pricing

LTP   Long-term planning

M&V   Measurement and verification

MQTT   Message Queuing Telemetry Transport

MW/MWh   Megawatt/megawatt hour

NEM   National Electricity Market

NIST   National Institute of Standards and Technology

NOC   Network operations center

NREL   National Renewable Energy Laboratory

NSA   U.S. National Security Agency

OCPP   Open Charge Point Protocol

OPC-UA   Ole for Process Control—Universal Architecture

OPEX Operational expenses

OSCP Open Smart Charging Protocol

OT Operational technology

PaaS Platform as a Service

PKI Public key infrastructure

PMU Phasor measurement unit

PNNL Pacific Northwest National Laboratory

PPA Power purchase agreement

pub/sub Publication/subscribe

PUC Public Utility Commission

PV Photovoltaic

REST Representational State Transfer

ROI Return on investment

SaaS Software as a Service

SCADA Supervisory Control and Data Acquisition

SEP Smart Energy Profile

SEPA Smart Electric Power Alliance

SGIP Smart Grid Interoperability Panel

SMCU Smart inverter control unit

SOA Service-oriented architecture

SSL Secure Socket Layer

STP Short-term planning

T-D Transmission–distribution

TLS Transport Layer Security

TSO Transmission system operator

TW/TWh Terawatt/terawatt hour

UML   Universal Markup Language

VAR   Volt-amps reactive

VPP   Virtual power plant

WMS   Workforce management system

WYSIWYG   What you see is what you get

XML   Extensible Markup Language

About the Author

Stuart McCafferty is a Cleanie Award-winning thought leader for his numerous articles on climate change and the electric power industry's opportunity to solve it using modern technologies, IoT architectures, clean energy systems, energy markets, and practical approaches. He genuinely believes that we can solve the climate change crisis, and the electric power industry is not only the most impactful opportunity, but the industry is also up for the challenge.

Mr. McCafferty is a United States Air Force Academy (USAFA) graduate, with a BS in engineering, and grew up as a Navy brat where he learned to make friends quickly. He has lived in many U.S. and global locations, but has spent the majority of his professional life in the high-tech city of Huntsville, Alabama. He has been fortunate to be surrounded by smart, technical, and creative colleagues throughout his career. Although he works tirelessly to tackle serious environmental and technical issues, he is remarkably lighthearted with a humorous and satirical view of just about everything.

He is a very active, but poor golfer, spending most weekends on courses with his wife, Cindy, and golfer buddies with the dream of living long enough to one day shoot his age. He and Cindy plan to retire outside of Augusta, Georgia, in the golfing and lake community of Lake Oconee. However, with all the exciting opportunities, incredible advances in technology, and problems that need to be fixed, full-time retirement is not in the plan.

Index

Synergies for Sustainable Energy, Elvin Yüzügüllü

A Systems Approach to Lithium-Ion Battery Management, Phil Weicker

Telecommunication Networks for the Smart Grid, Alberto Sendin,
Miguel A. Sanchez-Fornie, Iñigo Berganza, Javier Simon, and Iker Urrutia

A Whole-System Approach to High-Performance Green Buildings, David Strong
and Victoria Burrows

For further information on these and other Artech House titles, including previously considered out-of-print books now available through our In-Print-Forever® (IPF®) program, contact:

Artech House	Artech House
685 Canton Street	16 Sussex Street
Norwood, MA 02062	London SW1V 4RW UK
Phone: 781-769-9750	Phone: +44 (0)20 7596-8750
Fax: 781-769-6334	Fax: +44 (0)20 7630-0166
e-mail: artech@artechhouse.com	e-mail: artech-uk@artechhouse.com

Find us on the World Wide Web at: www.artechhouse.com